SAFE DRINKING WATER ACT AND ITS INTERPRETATION

SAFE DRINKING WATER ACT AND ITS INTERPRETATION

THOMAS W. CARTER
EDITOR

Novinka Books
New York

For permission to use material from this book please contact us:
Telephone 631-231-7269; Fax 631-231-8175
Web Site: http://www.novapublishers.com

Library of Congress Cataloging-in-Publication Data
Available upon request

ISBN 1-59454-761-0

Published by Nova Science Publishers, Inc. ✣New York

CONTENTS

PREFACE

Key drinking water issues include problems caused by specific contaminants, such as the gasoline additive methyl tertiary butyl ether (MTBE), perchlorate, and lead, as well as the related issue of the appropriate federal role in providing financial assistance for water infrastructure projects. Congress last reauthorized the Safe Drinking Water Act (SDWA) in 1996, and although funding authority for most SDWA programs expired in FY2003, broad reauthorization efforts are not expected as EPA, states, and water utilities remain busy implementing the requirements of the 1996 amendments. Concerns about perchlorate in drinking water also have returned to the congressional agenda, after the past Congress enacted several provisions on this issue. H.R. 213 has been introduced to require EPA to set a drinking water standard for perchlorate in 2007, and a January 2005 National Academy of Sciences report on the health effects of perchlorate has increased oversight interest in perchlorate regulatory activities at EPA. Concerns over the security of the nation's drinking water supplies were addressed by the 107th Congress through the Bioterrorism Preparedness Act (P.L. 107-188), which amended SDWA to require community water systems to conduct vulnerability assessments and prepare emergency response plans. Subsequent congressional action has involved oversight and funding of water security assessment and planning efforts and research. An ongoing SDWA issue involves the growing cost and complexity of drinking water standards and the ability of water systems, especially small, rural systems, to comply with standards. The issue of the cost of drinking water standards, particularly the new arsenic standard, has merged with the larger debate over the federal role in assisting communities with financing drinking water infrastructure - an issue that has become more challenging in a time of tightened budgets. Congress authorized a drinking water state revolving fund

(DWSRF) program in 1996 to help communities finance projects needed to meet standards. For FY2005, Congress provided $843 million for the DWSRF program, and the President has requested $850 million for FY2006. Notwithstanding this program, studies show that a large funding gap exists and will grow as SDWA requirements increase and infrastructure ages.

In: Safe Drinking Water Act and its Interpretation ISBN: 1-59454-761-0
Editor: Thomas W. Carter, pp. 1-23

Chapter 1

SAFE DRINKING WATER ACT: IMPLEMENTATION AND ISSUES*

Mary Tiemann

ABSTRACT

Key drinking water issues on the agenda in the 109th Congress include problems caused by specific contaminants, such as the gasoline additive methyl tertiary butyl ether (MTBE), perchlorate, and lead, as well as the related issue of the appropriate federal role in providing financial assistance for water infrastructure projects. Congress last reauthorized the

Safe Drinking Water Act (SDWA) in 1996, and although funding authority for most SDWA programs expired in FY2003, broad reauthorization efforts are not expected as EPA, states, and water utilities remain busy implementing the requirements of the 1996 amendments.

The 109th Congress is continuing efforts to address MTBE contamination of public water supplies. S. 606 and the House Energy and Commerce Committee discussion draft energy bill propose to strengthen the leak prevention provisions of the federal underground storage tank regulatory program and authorize funding from the Leaking Underground Storage Tank (LUST) Trust Fund for the cleanup of tank leaks involving MTBE.H.R. 789 would require secondary containment for tanks installed near public water supplies or private wells. Concerns about perchlorate in drinking water also have returned to the congressional agenda, after the past Congress enacted several provisions on this issue. H.R. 213 has been introduced to require EPA to set a drinking water standard for perchlorate

* Excerpted from CRS Report IB10118 dated March 15, 2005.

in 2007, and a January 2005 National Academy of Sciences report on the health effects of perchlorate has increased oversight interest in perchlorate regulatory activities at EPA.

Concerns over the security of the nation's drinking water supplies were addressed by the 107th Congress through the Bioterrorism Preparedness Act (P.L. 107-188), which amended SDWA to require community water systems to conduct vulnerability assessments and prepare emergency response plans. Subsequent congressional action has involved oversight and funding of water security assessment and planning efforts and research.

An ongoing SDWA issue involves the growing cost and complexity of drinking water standards and the ability of water systems, especially small, rural systems, to comply with standards. The issue of the cost of drinking water standards, particularly the new arsenic standard, has merged with the larger debate over the federal role in assisting communities with financing drinking water infrastructure — an issue that has become more challenging in a time of tightened budgets. Congress authorized a drinking water state revolving fund (DWSRF) program in 1996 to help communities finance projects needed to meet standards. For FY2005, Congress provided $843 million for the DWSRF program, and the President has requested $850 million for FY2006. Notwithstanding this program, studies show that a large funding gap exists and will grow as SDWA requirements increase and infrastructure ages.

In the past Congress, several bills were offered to increase funding for the DWSRF, provide more technical assistance to small systems, and/or create grant programs for qualified small systems; however, none of the bills was enacted. These issues remain on the agenda in the 109th Congress.

Most Recent Developments

The President's FY2006 budget request includes $850 million for the drinking water state revolving fund (DWSRF) program, $105.1 million for state public water system supervision (PWSS) grants, $11 million for state underground injection control (UIC) grants, $5 million for drinking water state homeland security grants, and $44 million for new water security initiatives. For FY2005 (in P.L. 108-447), the 108th Congress provided $850 million for the DWSRF program ($843 million, after applying a mandatory 0.8% across-the-board reduction to accounts funded in this act). Congress also provided $100.5 million for PWSS grants, $10.8 million for UIC grants, and $5 million for drinking water state homeland security grants, all subject to the mandatory 0.8% reduction. Conferees directed EPA to report, by August 2005, on the impact of the arsenic rule on communities, and to propose compliance alternatives and make recommendations to minimize compliance costs.

BACKGROUND AND ANALYSIS

Introduction

The Safe Drinking Water Act (SDWA), Title XIV of the Public Health Service Act (42 U.S.C. 300f-300j-26), is the key federal law for protecting public water supplies from harmful contaminants. First enacted in 1974 and widely amended in 1986 and 1996, the Act is administered through programs that regulate contaminants in public water supplies, provide funding for infrastructure projects, protect sources of drinking water, and promote the capacity of water systems to comply with SDWA regulations. The 1974 law established the current federal-state arrangement in which states and tribes may be delegated primary enforcement and implementation authority (primacy) for the drinking water program by the Environmental Protection Agency (EPA), which is the federal agency responsible for administering the law. The state-administered Public Water Supply Supervision (PWSS) Program remains the basic program for regulating public water systems, and EPA has delegated primacy for this program to all states, except Wyoming and the District of Columbia (which is defined as a state under SDWA); EPA has responsibility for implementing the PWSS program in these two jurisdictions. (See also CRS Report RL31243, *Safe Drinking Water Act: A Summary of the Act and Its Major Requirements*.)

More than 90% of people in the United States get their drinking water from one of the nearly 53,400 community water systems nationwide. Congress passed the SDWA in 1974, after a nationwide study of community water systems revealed widespread water quality problems and health risks resulting from poor operating procedures, inadequate facilities, and poor management of water supplies in communities of all sizes. Since then, government and private efforts to implement the Act have led to better public water system management and more information about, and greater confidence in, the quality of water provided at the tap.

Significant progress has been made during the 28 years of the federal drinking water program. Some 91 drinking water contaminants are now regulated, and EPA reports that the population served by community water systems that met all health-based standards increased from 83% in 1994 to 91% in 2002. Nonetheless, drinking water safety concerns and challenges remain. EPA and state enforcement data indicate that public water systems still incur tens of thousands of violations of SDWA requirements each year. These violations primarily involve monitoring and reporting requirements, but also include thousands of violations of standards and treatment techniques. Moreover, monitoring and reporting violations create uncertainty as to whether systems actually met the applicable health-based standards. Concern also exists over the potential health effects of contaminants for which drinking water standards have not been set, such as perchlorate and MTBE.

The 1996 SDWA Amendments

The 104th Congress made numerous changes to the Act with the SDWA Amendments of 1996 (P.L. 104-182), culminating a multi-year effort to amend a law that was widely criticized as having too little flexibility, too many unfunded mandates, and an arduous but unfocused regulatory schedule. Among the key provisions, the 1996 amendments authorized a drinking water state revolving loan fund (DWSRF) program to help public water systems finance projects needed to comply with SDWA rules. The amendments also established a process for selecting contaminants for regulation based on health risk and occurrence, gave EPA some added flexibility to consider costs and benefits in setting most new standards, and established schedules for regulating certain contaminants (such as *Cryptosporidium*, arsenic, and radon). The law added several provisions aimed at building the capacity of water systems (especially small systems) to comply with SDWA regulations, and it imposed many new requirements on the states including programs for source water assessment, operator certification and training, and compliance capacity development. The amendments also required that community water suppliers provide customers with annual "consumer confidence reports" that provide information on contaminants found in the local drinking water. The law authorized appropriations for SDWA programs through FY2003.

Regulated Public Water Systems

Federal drinking water regulations apply to some161,000 privately and publicly owned water systems that provide piped water for human consumption to at least 15 service connections or that regularly serve at least

25 people. (The law does not apply to private, residential wells.) Of these systems, roughly 53,400 are *community water systems* (CWSs) that serve a residential population of nearly 270 million year-round. All federal regulations apply to these systems. (Roughly 15% of community systems are investor-owned.) Nearly 18,700 public water systems are *non-transient, non-community water systems* (NTNCWSs), such as schools or factories, that have their own water supply and serve the same people for more than six months but not year-round. Most drinking water requirements apply to these systems. Another 89,000 systems are *transient non-community water systems* (TNCWSs) (e.g., campgrounds and gas stations) that provide their own water to transitory customers. TNCWSs generally are required to comply only with regulations for contaminants that pose immediate health risks (such as microbial contaminants), with the proviso that systems that use surface water sources must also comply with filtration and disinfection regulations.

Of the 53,363 community water systems, roughly 84% serve 3,300 or fewer people. While large in number, these systems provide water to just 10% of the population served by all community systems. In contrast, 7% of community water systems serve more than 10,000 people, and they provide water to 81% of the population served. Fully 85% (15,900) of non-transient, non-community water systems and 97% (86,400) of transient noncommunity water systems serve 500 or fewer people. These statistics give some insight into the scope of financial, technological, and managerial challenges many public water systems face in meeting a growing number of complex federal drinking water regulations. **Table 1** provides statistics for community water systems.

CURRENT DRINKING WATER ISSUES

Major drinking water issues involve infrastructure funding needs; the security of water supplies; small system capacity to comply with SDWA; and contamination of drinking water by specific contaminants, including lead and the unregulated contaminants, MTBE and perchlorate. Although appropriations for most SDWA programs were authorized through FY2003, SDWA reauthorization was not on the agenda in the 108th Congress. Rather, various bills were offered to address specific issues, such as infrastructure funding and contamination by lead, MTBE, and perchlorate. As with other EPA-administered statutes having expired funding authority, the programs do not expire as long as Congress continues to appropriate funds for these programs. (For information on water supply issues and legislation, see CRS Issue Brief IB10019, *Western Water Resource Issues.*)

Table 1. Size Categories of Community Water Systems

System size (population served)	Number of Community Water Systems	Population Served (millions)	Percent of Community Water Systems	Percent of Population Served
Very small (25-500)	30,417	5.01	57%	2%
Small (501-3,300)	14,394	20.26	27%	7%
Medium (3,301-10,000)	4,686	27.20	9%	10%
Large (10,001-100,000)	3,505	98.71	7%	36%
Very large (>100,000)	361	122.15	1%	45%
Total	53,363	273.33	100%	100%

Adapted from: US Environmental Protection Agency. Factoids: Drinking Water and Ground Water Statistics for 2003. Available at Internet website: [http://www.epa.gov/safewater/data/pdfs/factoids_2003.pdf].

Regulating Drinking Water Contaminants

Standard-Setting

The Safe Drinking Water Act directs EPA to promulgate National Primary Drinking Water Regulations for contaminants that may pose public health risks and that are likely to be present in public water supplies. These regulations generally include numerical standards to limit the amount of a contaminant that may be present in drinking water. Where it is not economically and technically feasible to measure a contaminant at very low concentrations, EPA establishes a treatment technique in lieu of a standard.

To develop a drinking water regulation, EPA must address a variety of technical issues. The agency must (1) determine the occurrence of a contaminant in the environment, and especially in public water systems; (2) evaluate human exposure and risks of adverse health effects to the general population and to sensitive subpopulations; (3) ensure that analytical methods are available for water systems to use in monitoring for a contaminant; (4) evaluate the availability and costs of treatment techniques that can be used to remove a contaminant; and (5) assess the impacts of a regulation on public water systems, the economy, and public health. Consequently, regulation development typically is a multi-year process. EPA may expedite procedures and issue interim standards to respond to urgent threats to public health.

After reviewing health effects studies, EPA sets a nonenforceable maximum contaminant level goal (MCLG) at a level at which no known or anticipated adverse health effects occur and that allows an adequate margin of safety. EPA also considers the risk to sensitive subpopulations (e.g., children). For carcinogens and microbes, EPA sets the MCLG at zero. Because MCLGs consider only health effects and not analytical detection limits or treatment technologies, they may be set at levels that water systems cannot meet. Once the MCLG is established, EPA sets an enforceable standard, the maximum contaminant level (MCL). The MCL generally must be set as close to the MCLG as is "feasible" using the best technology or other means available, taking costs into consideration.

EPA has relied on legislative history to determine the meaning of "feasible." Most recently, the Senate report accompanying the 1996 amendments stated that feasible means the level that can be reached by large, regional drinking water systems applying best available treatment technology. The report explained that this approach is used because 80% of the population receives its drinking water from large community water systems, and thus, safe water can be provided to most of the population at very affordable costs. (About 80% of the population is served by systems that serve a population of 10,000 or more.) However, because standards are based on cost considerations for large systems, Congress expected that standards could be less affordable for smaller systems. An issue in the 1996 reauthorization debate concerned whether the costs of some standards were justified, given their estimated risk-reduction benefits. As amended, the Act now requires EPA, when proposing a standard, to publish a determination as to whether or not the benefits of a proposed standard justify the costs. If EPA determines that the benefits do not justify the costs, EPA, in certain cases, may promulgate a standard less stringent than the feasible level that "maximizes health risk reduction benefits at a cost that is justified by the benefits."

Recent and Pending Rules

EPA's recent rulemaking activities include a 1998 rule package that expanded requirements to control pathogens, especially *Cryptosporidium* (Interim Enhanced Surface Water Treatment Rule (SWTR)) and disinfectants (e.g., chlorine) and their byproducts (e.g., chloroform) (Stage 1 Disinfectant and Disinfection Byproduct Rule). In 2002, EPA issued the Long Term 1 Enhanced SWTR to improve control of microbial pathogens among small systems. EPA also has issued new rules for several radionuclides, including radium (now in effect), and a revised standard for arsenic that water systems must comply with by January 23, 2006.

EPA has nearly completed several related rulemakings, including a groundwater rule to establish disinfection requirements for systems relying on ground water (this rule is intended to protect against fecal bacteria contamination in these systems); and a rule package (expected in July 2005) that includes the Stage 2 Disinfectants and Disinfection Byproduct Rule and the Long Term 2 Enhanced Surface Water Treatment Rule. These rules build on the rules issued in 1998 to strengthen public health protection from disinfectants, their byproducts, and pathogens. EPA also is working to issue a radon rule, and is evaluating many other contaminants, including perchlorate and MTBE, for possible regulation.

Perchlorate

Perchlorate is the main ingredient of solid rocket fuel and has been used heavily by the Department of Defense (DOD), the National Aeronautics and Space Administration (NASA), and related industries. This highly soluble and persistent compound has been disposed of on the ground for decades, and now has been detected in sources of drinking water that serve more than 11 million people. Perchlorate is known to disrupt the uptake of iodine in the thyroid; thus, perchlorate can affect thyroid function. A key concern is that, if sufficiently severe, impaired thyroid function in pregnant women can impair brain development in fetuses and infants.

EPA identified perchlorate as a candidate for regulation in 1998, but concluded that information was insufficient at that time to make a regulatory determination. EPA listed perchlorate as a priority for further research on health effects and treatment technologies, and for collecting occurrence data. In 2002, EPA issued a controversial draft risk assessment for perchlorate that concluded that potential human health risks of perchlorate exposure include effects on the developing nervous systems and thyroid tumors, based on rat studies that observed benign tumors and adverse effects in fetal brain development. The draft assessment included a revised draft reference dose (RfD) intended to protect the most sensitive groups against these effects. That dose roughly translated to a drinking water standard of 1 part per billion (ppb). EPA's 1999 draft level translated to a standard of roughly 32 ppb.

Because an RfD provides the basis for determining the level at which a standard is set, and because drinking water standards are often used as environmental cleanup standards, the DOD and other major perchlorate users have followed EPA's efforts closely. Interagency debate over the draft assessment persisted, and in March 2003, EPA, the DOD, NASA, and other federal agencies asked the National Research Council (NRC) of the National Academy of Sciences to review the science for perchlorate and EPA's draft risk assessment.

The NRC released its study in January 2005, and broadly agreed with several EPA findings; however, the NRC committee suggested several changes to EPA's draft risk assessment. Among other findings, the committee noted that, unlike rats, humans have multiple mechanisms to compensate for iodide deficiency and thyroid disorders, and that studies in rats are of limited use for quantitatively assessing human health risk associated with perchlorate exposure. The committee recommended that EPA base its assessment on human data. The NRC calculated an RfD for perchlorate that incorporates an uncertainty factor to protect the most sensitive populations; that RfD would translate to a drinking water equivalent level of 24.5 ppb. (In developing an MCLG, EPA would likely lower this number to reflect the amount of perchlorate exposure that EPA determines comes from other sources, especially food.) In February, EPA adopted the NRC's recommended reference dose. (For further discussion, see CRS Report RS21961, *Perchlorate Contamination of Drinking Water: Regulatory Issues and Legislative Actions*.)

Congressional interest in this issue continues, and perchlorate legislation has been offered again in the 109th Congress. H.R. 213 would require EPA to promulgate a drinking water standard for perchlorate by July 31, 2007. The 108th Congress passed several perchlorate measures. The Department of Defense Authorization Act of FY2004 (P.L. 108-136) required DOD to provide for health studies of perchlorate in drinking water. The DOD FY2004 Appropriations Act (P.L. 108-87) directed DOD, with EPA, to study perchlorate groundwater pollution that threatens drinking water and irrigation supplies in the Southwest. The National Defense Authorization Act for FY2005 (P.L. 108-375) included a "Sense of Congress" that DOD should develop a plan for remediating perchlorate contamination resulting from DOD activities to ensure DOD can respond quickly once a federal drinking water standard is established; continue remediating sites where perchlorate contamination poses an imminent and substantial endangerment to human health and welfare; develop a plan to remediate contamination when the Secretary determines that the contamination poses a health hazard; and continue evaluating sites, even in the absence of an SDWA standard. Also, S. 2550 (S.Rept. 108-386), a water infrastructure bill, would have required the U.S. Geological Survey to conduct a national survey on perchlorate contamination. Several other bills would have required EPA to promulgate a drinking water standard for perchlorate.

Lead in Drinking Water

Lead from various sources (including paint in older homes, soil, and water) poses one of the main environmental threats to children's health. In 2004, the issue of lead contamination reemerged in Washington, D.C., where water monitoring revealed marked increases in the levels of lead in tap water in recent years. The local water authority's limited response to the monitoring results severely damaged public trust in the local water supply. These events led policy makers and EPA to examine the effectiveness of the lead rule, particularly its monitoring and public notification requirements, as well as compliance with the regulation. In the 108th Congress, hearings were held by the House Energy and Commerce Committee (July 22, 2004), the House Government Reform Committee (March 5 and May 21, 2004), and the Senate Environment and Public Works Committee (April 7, 2004). H.R. 4268 and S. 2377 were introduced to strengthen the regulation of lead in drinking water and to remediate lead in school drinking water. S. 2550, a water infrastructure financing bill, also included lead provisions.

Lead Rule Overview

In 1991, EPA issued the Lead and Copper Rule (56 *FR* 26460) to replace an interim lead regulation that included a standard of 50 parts per billion (ppb) that was outdated and not protective of public health. Epidemiological research had shown that adverse health effects from exposures to lead occur at lower levels and are worse than previously thought, particularly for infants and children. (There is no known safe level of exposure to lead, and recent studies suggest that very low levels of lead may adversely affect children's neurological development.) In 1988, EPA had proposed a regulation that would have established an enforceable lead standard (maximum contaminant level (MCL)) of 5 ppb applicable to water leaving the treatment plant and also would have required a treatment technique (corrosion control) to further reduce lead in drinking water. Commenters on the proposal expressed concern that a standard applicable at the treatment plant would not indicate the amount of lead in tap water, and that compliance at the tap was essential. EPA and utilities were concerned that an MCL applied at the tap would not be feasible because lead in household plumbing could be a major cause of violations of a lead standard applied at the tap — a situation beyond the control of the water system. This issue reflected the problem with regulating lead. Unlike most contaminants, lead is not normally present in water as it leaves the treatment plant; rather, lead occurs in drinking water primarily as a corrosion by-product, entering water as it travels through pipes in the distribution system and in household plumbing. The primary sources of lead in drinking water are lead pipes, lead solder that has been used in plumbing systems, and brass plumbing fixtures that contain lead.

The final Lead and Copper Rule (LCR) did not include an enforceable standard. Instead, the LCR established a treatment technique (corrosion control) to prevent lead from leaching into drinking water. (Optimizing corrosion control is a complex process, and the "optimal" treatment can change as water characteristics change and as utilities add new treatment processes to meet other drinking water regulations.)

The lead rule established a lead "action level" of 15 ppb at the tap, based on the 90th percentile level of water samples. Water systems are required to sample tap water in homes and buildings that are at high risk of lead contamination. If lead concentrations exceed 15 ppb in more than 10% of taps sampled, the system is required, within 60 days, to inform customers about lead's health effects and sources, and what can be done to reduce exposure. The system must continue to deliver educational materials as long as it exceeds the action level. If the system continues to exceed the action level after installing optimal corrosion control, it must replace 7% of the lead service lines under its ownership each year, and must offer to replace the privately owned portion of a service line (at the owner's expense).

Federal and Local Efforts

EPA, the D.C. Water and Sewer Authority (WASA), and other local officials worked with the U.S. Army Corps of Engineers to determine the cause of the elevated lead levels in the District of Columbia. (The Corps treats and supplies water from the Washington Aqueduct to the District and several communities.) It appears that changes in treatment processes, made by the Corps to comply with another EPA regulation, made the water more corrosive, thus causing more lead to be leached from lead pipes in the distribution system and from lead plumbing inside homes and other buildings. In late 2000, the Corps changed the chemicals in its secondary disinfection treatment from free chlorine to chloramines to comply with an EPA regulation that placed strict limits on disinfection byproducts. Starting with the monitoring period, July 2001 through June 2002, more than 10% of tap water samples taken by WASA exceeded the lead action level.

The Corps of Engineers began testing a new corrosion control treatment process in June 2004. In August 2004, EPA approved use of the process for the entire Aqueduct service area and imposed supplemental monitoring and reporting requirements on the affected public water systems. Additionally, EPA determined that WASA had failed to comply with numerous lead sampling, public notification, and reporting requirements. EPA and WASA reached a consent agreement that requires WASA to replace more than 1,600 lead service lines, improve its public education program, and upgrade its database management systems.

National Review

EPA has undertaken a national review of lead monitoring by water systems since 2000 to determine whether the lead problem in the District was widespread. By June 2004, EPA had received monitoring data for 744 (89%) of the 834 systems that serve more than 50,000 people. EPA reported that 27 of these systems (3.6%) exceeded the action level at least once since 2000, and 12 of the systems exceeded the action level during 2003. Most (66%) of the systems serving more than 50,000 people reported that the highest level observed during any monitoring period since 2000 was less than 5 ppb. For systems serving between 3,300 and 50,000 people, 237 (3.4%) of 7,833 systems reporting had exceeded the action level since 2000; 76 systems exceeded the action level for monitoring period ending after January 2003. Most systems (71%) in this size category reported that the highest level of lead observed since 2000 was less than 5 ppb. In October 2004, EPA announced that the national data from 73,000 water utilities demonstrated that lead in drinking water is not a widespread problem.

EPA also has been assessing national compliance with the lead rule and reviewing the rule to determine whether major changes are needed. Parts of the regulation that are receiving most scrutiny include the public notification, monitoring, and lead service line replacement requirements. In November, EPA issued a guidance memo to clarify sampling requirements. In March 2005, EPA initiated a Drinking Water Lead Reduction Plan, based on its review of the lead rule. Under the plan, EPA intends to tighten and clarify monitoring and public notification requirements, and revise treatment and lead service line replacement requirements. EPA plans to propose changes to the lead rule by early 2006. EPA also will revise its 1994 guidance on testing for lead in school drinking water. (See also [http://www.epa.gov/safewater/lcrmr/lead_review.html] and CRS Report RS21831, *Lead in Drinking Water: Washington, D.C., Issue and Broader Regulatory Implications*.)

Methyl Tertiary Butyl Ether (MTBE)

For technological and cost reasons, this gasoline additive has been widely used to meet the Clean Air Act requirement that reformulated gasoline (RFG) contain at least 2% oxygen to improve combustion. RFG is required for use in areas that fail to meet the federal ozone standard. However, numerous incidents of water contamination by MTBE have led to calls for restrictions on its use. Nineteen states, including California and New York, have enacted limits or phase-outs of the additive. EPA has not developed a drinking water standard for MTBE; however, at least 7 states have set their own MTBE drinking water standard.

The primary source of MTBE in drinking water has been petroleum releases from leaking underground storage tank (UST) systems. Once released, MTBE moves through soil and into water more rapidly than other gasoline components, thus making it is more difficult and costly to clean up than conventional gasoline leaks. Although MTBE is considered to be less toxic than some other gasoline components (such as benzene), even small amounts of MTBE can render water undrinkable because of its strong taste and odor. These characteristics have made MTBE use an important issue for water suppliers and consumers.

In 1997, EPA issued a drinking water advisory for MTBE based on consumer acceptability (for taste and smell). Advisories provide information on contaminants that have not been regulated under SDWA. They are not enforceable, but provide guidance to water suppliers and others regarding potential health effects or consumer acceptability. While the MTBE advisory is not based on health effects, EPA states that keeping MTBE levels in the range of 20-40 micrograms per liter (:g/L) or lower for consumer acceptability reasons would also provide a large margin of safety from potential adverse health effects.

EPA has taken steps that could lead to the issuance of a drinking water standard for MTBE. In 1998, EPA included MTBE on a list of contaminants that are potential candidates for regulation. Compounds on the contaminant candidate list are categorized as regulatory determination priorities, research priorities, or occurrence priorities. Because of data gaps on health effects and occurrence, EPA placed MTBE in the category of contaminants for which further occurrence data collection and health effects research are priorities. Thus, although EPA did not select MTBE for regulation, the agency has pursued research to fill data gaps so that a regulatory determination may be made. The next round of determinations is scheduled for 2006, although EPA can make determinations outside of this cycle.

The 108th Congress passed several bills that addressed drinking water contamination by MTBE, but none was enacted. The Senate passed an underground storage tank bill, S. 195 (S.Rept. 108-13), which would have authorized appropriations from the Leaking Underground Storage Tank (LUST) Trust Fund for cleaning up MTBE contamination and would have added new leak prevention, inspection, and enforcement requirements to the federal tank regulatory program. The comprehensive energy bill, H.R. 6 (H.Rept. 108-375) included a similar range of UST regulatory provisions (but with key differences, including less frequent tank inspection requirements) and authorized trust fund appropriations to address leaks containing MTBE or other oxygenated fuel additives, such as ethanol.

The conference report for H.R. 6 also included a contentious "safe harbor" provision to prohibit products liability lawsuits, alleging manufacturing or design defects, against producers of fuels containing MTBE and renewable fuels, such as ethanol and bio-diesel. The safe harbor provision would not affect liability for remediation costs, drinking water contamination, or negligence; however, with liability for manufacturing and design defects ruled out, plaintiffs would have to demonstrate negligence in the handling of such fuels, a more difficult legal standard to meet. Public water suppliers, and state and local government associations, strongly oppose a safe harbor provision and express concern that it could leave communities paying much of the cost for cleaning up contamination by fuels containing MTBE or renewable fuels. Manufacturers argue that a safe harbor provision is reasonable, given that MTBE has been used to meet federal mandates, and that the key problem lies with leaking tanks, not with MTBE. The House passed the conference report, but a cloture vote failed in the Senate. No further action occurred on this bill.

In the 109th Congress, the House Energy and Commerce Committee has offered a discussion draft energy bill that is essentially the same as H.R. 6. In the Senate, S. 606 includes similar LUST and MTBE provisions, but excludes the safe harbor for MTBE. (For more information on these issues, see also CRS Report RL32787, *MTBE in Gasoline: Clean Air and Drinking Water Issues*.)

Drinking Water Infrastructure Funding

Drinking Water State Revolving Fund

A persistent SDWA issue concerns the ability of public water systems to upgrade or replace infrastructure to comply with federal drinking water regulations and, more broadly, to ensure the provision of a safe and reliable water supply. In the 1996 SDWA Amendments, Congress responded to growing complaints about the Act's unfunded mandates and authorized a drinking water state revolving loan fund (DWSRF) program to help water systems finance infrastructure projects needed to meet drinking water standards and address the most serious health risks. The program authorizes EPA to award annual capitalization grants to states. States then use their grants (plus a 20% state match) to provide loans and other assistance to systems. Communities repay loans into the fund, thus making resources available for projects in other communities. Eligible projects include installation and replacement of treatment facilities, distribution systems, and certain storage facilities. Projects to replace aging infrastructure are eligible if they are needed to maintain compliance or to further public health protection goals.

Congress authorized funding totaling $9.6 billion, including $1 billion for each of FY1995 through FY2003 for the DWSRF program. To date, Congress has provided roughly $7.8 billion for this program, including $843 million in the FY2005 omnibus spending bill, P.L. 108-447. For FY2006, the President has requested $850 million. Through June 2004, EPA had awarded $5.74 billion in capitalization grants, which, when combined with the state match, bond proceeds, and other funds, amounted to $9.64 billion in DWSRF funds available for loans and other assistance. Through that same period, 6,500 drinking water system projects had received assistance, and total assistance provided by the program reached $7.98 billion. (For further information, see CRS Report RS22037, *Drinking Water State Revolving Fund: Program Overview and Issues*.)

Funding Issues

The DWSRF program is generally well regarded; however, many organizations and state and local officials argue that greater investment in drinking water infrastructure is needed. EPA's latest survey of capital improvement needs for water systems estimated that communities need to invest $150.9 billion on drinking water infrastructure improvements over 20 years (1999-2018) to comply with existing drinking water regulations and to ensure the provision of safe water. The survey excluded funds needed for compliance with several recent regulations (including the revised arsenic and radium rules) and pending rules for radon and other contaminants; nor did it consider funds needed for security upgrades. These requirements are expected to substantially increase needs estimates.

A related issue is the need for communities to address infrastructure needs that are outside the scope of the DWSRF program and, thus, generally are ineligible for assistance from this source. Ineligible categories include future growth, ongoing rehabilitation, and operation and maintenance of systems. According to EPA, outdated and deteriorated drinking water infrastructure poses a fundamental long-term threat to drinking water safety, and in many communities, basic infrastructure costs could far exceed SDWA compliance costs.

In 2002, EPA issued *The Clean Water And Drinking Water Infrastructure Gap Analysis* which identified potential funding gaps between projected needs and spending from 2000 through 2019. This analysis estimated the potential 20-year funding gap for drinking water and wastewater infrastructure capital and operations and maintenance (O&M), based on two scenarios: a "no revenue growth" scenario and a "revenue growth" scenario that assumed spending on infrastructure would increase 3% per year. Under the "no revenue growth" scenario, EPA projected a funding gap for drinking water capital investment of $102 billion (roughly $5 billion per year) and an O&M funding gap of $161 billion ($8 billion per year). Using revenue growth assumptions, EPA estimated a 20-year capital funding gap of $45 billion ($2 billion per year), and no gap for O&M.

Other needs assessments also reveal a funding gap. A Congressional Budget Office study, *Future Investment in Drinking Water and Wastewater Infrastructure*, concluded that current funding from all levels of government, combined with current revenues from ratepayers, will not be sufficient to meet the nation's future demand for water infrastructure. In 2000, the Water Infrastructure Network (WIN) (a coalition of state and local officials, water service providers, environmental groups and others) reported that, over the next 20 years, water and wastewater systems need to invest $23 billion annually more than current investments to meet SDWA and Clean Water Act health and environmental priorities and to replace aging infrastructure. WIN and other groups have presented proposals to Congress for multi-billion dollar investment programs for water infrastructure. Others, however, have called for more financial self-reliance within the water sector.

In response to EPA's Gap Analysis, EPA's budget request for FY2004 proposed that funding for the DWSRF program be continued at a level of $850 million annually through FY2018. EPA's budget justification explained that this funding level would allow DWSRFs to revolve at a cumulative level of $1.2 billion (more than double the previous goal of $500 million) and would help close the funding gap for drinking water infrastructure needs.

In the face of large needs, tight budgets, and debate over the federal role in funding water infrastructure, EPA, states, and utilities have been examining alternative management and financing strategies to address costs. Strategies include establishing public-private partnerships (privatization options range from contracting for services to selling system assets), improving asset management, and adopting full-cost pricing for water services.

In the 108th Congress, several bills were introduced to increase DWSRF funding levels. S. 2550 (S.Rept. 108-386), a water infrastructure financing bill, would have authorized $15 billion over five years for the DWSRF and required states to reserve a portion of their DWSRF grant to make grants for up to 55% of project costs to qualified communities. The committee adopted various amendments, including a contentious provision that would have applied Davis-Bacon prevailing wage requirements, in perpetuity, to projects receiving DWSRF assistance. (For details, see CRS Report RL32503, *Water Infrastructure Financing Legislation: Comparison of S. 2550 and H.R. 1560.*)

Drinking Water Security

Congress addressed drinking water security issues in the Bioterrorism Preparedness of 2002 (P.L. 107-188, H.Rept. 107-481), which amended SDWA to require community water systems to conduct vulnerability assessments and prepare emergency response plans. In the 108th Congress, attention focused on several issues including the progress utilities have made in meeting the requirements of the Bioterrorism Act and in addressing identified vulnerabilities, and whether utilities need more resources to make security improvements. S. 2269 would have authorized EPA to make grants to utilities to improve security and authorized funds for the Water Information Sharing and Analysis Center (Water ISAC).

A key provision of the Bioterrorism Act required each community water system serving more than 3,300 individuals to assess their vulnerability to terrorist attacks or other intentional acts to disrupt the provision of a safe and reliable drinking water supply. Combined, these systems serve more than 90% of the population served by community water systems. The Act required these systems to certify to EPA that they conducted a vulnerability assessment and to give EPA a copy of the assessment. The Act also required these systems to prepare or revise emergency response plans incorporating the results of the vulnerability assessments no later than six months after completing them. Table 2 outlines the schedule for utilities to submit their assessments to EPA and to complete emergency response plans.

Table 2. Community Water System Requirements under the Bioterrorism Act

System size by population (approx. no. of systems)	Vulnerability assessments must be completed (% completed as of 10/1/04)	Emergency response plans must be completed (% certified as of 10/1/04)
100,000 or more (425)	March 31, 2003 (100%)	September 30, 2003 (100%)
50,000 - 99,999 (460)	December 31, 2003 (98%)	June 30, 2004 (99%)
3,301 - 49,999 (7,500)	June 30, 2004 (88%)	December 31, 2004 (NA)

The Bioterrorism Act authorized $160 million for FY2002, and sums as may be needed for FY2003 through FY2005, to provide financial assistance to community water systems to assess vulnerabilities, prepare response plans, and address security enhancements and significant threats. The emergency supplemental appropriations for FY2002 (P.L. 107-117) provided $90 million for assessing the vulnerabilities of drinking water utilities and other security planning, and $5 million for state grants for assessing drinking water safety. In FY2002, EPA awarded roughly $53 million in water security grants to help the largest public water systems complete vulnerability assessments by the March 31, 2003 deadline.

Federal grants were not available for smaller systems covered by the Bioterrorism Act's requirements. Instead, EPA, states and water organizations have provided vulnerability assessment tools, guidance documents, training, and technical assistance to support security enhancement efforts among these systems. Similar assistance is also being provided for remaining 84% of community water systems that serve 3,300 or fewer and are not required to do vulnerability assessments and emergency planning.

For FY2003, EPA requested $16.9 million for vulnerability assessments for small and medium-sized systems and $5 million for state water security coordinators to work with EPA and utilities in assessing water security. P.L. 108-7 included this amount, plus $2 million for the National Rural Water Association to help small systems with vulnerability assessments, and $1 million to the American Water Works Association to provide security training.

For FY2004, EPA requested and received $32.4 million for critical water infrastructure protection, including $5 million for state water security coordination grants. This funding supported states' efforts to work with water and wastewater systems to develop and enhance emergency operations plans; conduct training in the implementation of remedial plans in small systems; and develop detection, monitoring and treatment technology to enhance water security. EPA used funds to assist the nearly 8,000 community water systems that serve water to populations between 3,300 and 100,000 and are subject to the Bioterrorism Act.

For FY2005, EPA requested $5 million for state water security coordination grants and $6.1 million for other critical infrastructure protection efforts. EPA's budget justification explained that the $21.3 million reduction reflected a shift in priorities from assistance and training on vulnerability assessments. The Consolidated Appropriations Act for FY2005 provided this amount, including $2 million for the Water Information Sharing and Analysis Center, which shares sensitive security information with water systems.

In the FY2006 budget request, the President again has requested $5 million for state water security grants. The President also has requested $44 million to launch two new drinking water security initiatives, the Water Sentinel and the Water Alliance for Threat Reduction, in response to EPA's water security responsibilities under Homeland Security Presidential Directive (HSPD) 7, which designated EPA as the lead agency for water infrastructure security. The goal of the Water Sentinel initiative is to establish pilot early warning systems through intensive water monitoring and surveillance for certain chemical and biological contaminants in five cities, and to form a water laboratory alliance to build the analytical capacity needed to support the surveillance program. Under the Water Alliance for Threat Reduction initiative, EPA will work to ensure that large systems have the tools and information needed to prevent, detect, and respond to attacks. (See also CRS Report RL31294, *Safeguarding the Nation's Drinking Water: EPA and Congressional Actions*.)

Small Systems Issues

A key SDWA issue involves the financial, technical, and managerial capacity of small systems to comply with SDWA regulations. Roughly 84% of the nation's community water systems are small, serving 3,300 persons or fewer; 57% of the systems serve 500 persons or fewer. EPA and states have documented the problems many small systems face in meeting SDWA rules, and more fundamentally, in ensuring the quality of water supplies. Major problems include deteriorated infrastructure; lack of access to capital; limited customer and rate base; inadequate rates; diseconomies of scale; and limited technical and managerial capabilities. Although these systems serve just 9% of the population served by community water systems, the sheer number of small systems creates challenges for policymakers.

In the earliest SDWA debates, Congress recognized that setting standards based on technologies that are affordable for large cities could pose problems for small systems. During the reauthorization debate leading up to the 1996 amendments, policymakers gave considerable attention to the question of how to help small systems improve their capacity to ensure consistent compliance with the SDWA. The 1996 amendments added provisions aimed at achieving this goal, including a requirement that states establish strategies to assist systems in developing and maintaining the technical, financial and managerial capacity to meet SDWA regulations. Congress also revised provisions on standard-setting, variances, and exemptions to increase consideration of small system concerns.

Small System Variances

As amended in 1996, the SDWA requires EPA, when issuing a regulation, to identify technologies that meet the standard and that are affordable for systems that serve populations of 10,000 or fewer. If EPA does not identify "compliance" technologies that are affordable for these systems, then EPA must identify small system "variance" technologies. A variance technology need not meet the standard, but must protect public health. States may grant variances to systems serving 3,300 persons or fewer, if a system cannot afford to comply with a rule (through treatment, an alternative source of water, or other restructuring) and the system installs a variance technology. With EPA approval, states also may grant variances to systems serving between 3,300 and 10,000 people.

To date, EPA has determined that affordable compliance technologies are available for all drinking water regulations. Consequently, the agency has not identified any small system variance technologies, and *no small system variances are available*. If EPA had identified variance technologies, states still might not make much use of these variances for a number of reasons — a key issue being the creation of a double standard for tap water quality in communities that meet a standard, compared with those that would rely on variances.

Exemptions

The Act's exemption provisions also are intended to provide compliance flexibility in certain cases. States or EPA may grant temporary exemptions from a standard if, due to certain compelling factors (including cost), a system cannot comply on time. For example, all systems are required to comply with the new arsenic standard five years after its promulgation date. An exemption would allow three more years for qualified systems. Small systems (serving 3,300 persons or fewer) may be eligible for up to three additional two-year extensions, for a total exemption duration of nine years (for a total of up to 14 years to achieve compliance). In the preamble to the arsenic rule published in January 2001, EPA noted that exemptions will be an important tool to help states address the number of systems needing financial assistance to comply with this rule and other SDWA rules (66 *FR* 6988).

Affordability Issues and Arsenic Compliance

Prompted by intense debate over the revised arsenic standard and its potential cost to small communities, the conference report for EPA's FY2002 appropriations (H.Rept. 107-272) directed EPA to review its affordability criteria and how small system variance and exemption programs should be implemented for arsenic. Congress directed EPA to report on its affordability criteria, administrative actions, potential funding mechanisms for small system compliance, and possible legislative actions.

EPA's report to Congress, *Small Systems Arsenic Implementation Issues,* summarized activities that addressed these directives. Major activities included (1) reviewing the small system affordability criteria and variance process; (2) developing a small community assistance plan to improve access to financial and technical assistance, improve compliance capacity, and simplify the use of exemptions; and (3) implementing a $20 million research and technical assistance strategy. EPA has completed several efforts to help states and water systems meet the requirements of the arsenic rule. In August 2002, EPA issued *Implementation Guidance for the Arsenic Rule,* which includes guidance to help states grant exemptions. EPA has offered technical assistance and training to small systems, and is sponsoring research on low-cost treatment technologies for removing arsenic from drinking water. Also, EPA is working with small communities to maximize loans and grants under SDWA and the U.S. Department of Agriculture water infrastructure programs.

Water systems must comply with the new arsenic standard by January 23, 2006, and Congress has shown ongoing concern about compliance costs.

The conference report for the Consolidated Appropriations Act for FY2005 directs EPA to report, by August 2005, on the extent to which communities will be impacted by the arsenic rule, and to propose compliance alternatives and make recommendations to minimize compliance costs. Congress also provided $8.3 million for research on cost-effective arsenic removal technologies, which could reduce compliance costs. In the 109th Congress, S. 41 has been introduced to require states to grant small community water systems exemptions from regulations for naturally occurring contaminants in certain cases; a similar bill was offered in the 108th Congress. Also, in the past Congress, H.R. 3328/S. 1432 and S. 2550 all proposed to establish a small system grant program to help qualified communities comply with drinking water standards.

LEGISLATION

H.R. 213 (Solis)

The Safe Drinking Water for Healthy Communities Act of 2005 amends SDWA to require EPA to promulgate a drinking water standard for perchlorate by July 31, 2007. Introduced Jan. 4, 2005; referred to the Committee on Energy and Commerce.

H.R. 879 (Dingell)

Amends the Solid Waste Disposal Act to require secondary containment for all new and replaced underground storage tank systems located near public water systems and private drinking water wells to prevent contamination by petroleum and MTBE. Introduced on Feb. 17, 2005; referred to the Committee on Energy and Commerce.

S. 41 (Nelson, E. Benjamin)

Amends SDWA to direct states to grant small, nonprofit water systems exemptions from drinking water regulations for naturally occurring contaminants, in certain cases. Introduced Jan. 24, 2005; referred to the Committee on Environment and Public Works.

CONGRESSIONAL HEARINGS, REPORTS, AND DOCUMENTS

U.S. Congress. House. Committee on Energy and Commerce. Subcommittee on Environment and Hazardous Materials. *Tapped Out: Lead in the District of Columbia and the Providing of Safe Drinking Water.* Hearing, July 22, 2004, 108th Congress, 2nd session. 155 p. (108-97)

—— *Drinking Water Needs and Infrastructure.* Hearing, April 11, 2002. 107th Congress, 2nd session. 108 p. (107-107)

U.S. Congress. House. Committee on Government Reform. Subcommittee on Energy Policy, Natural Resources and Regulatory Affairs. *EPA Water Enforcement: Are We on the Right Track?* Hearing, Oct. 14, 2003, 108th Congress, 1st session. 201p. (108-157)

U.S. Congress. House. Committee on Government Reform. Public Confidence, Down the Drain: the Federal Role in Ensuring Safe Drinking Water in the District of Columbia.

Hearing, March 5, 2004, 108th Congress, 2nd session. 268 p. (108-161)

U.S. Congress. Senate. Committee on Environment and Public Works. *Water Infrastructure Financing Act.* Report to accompany S. 2550. Oct.7, 2004. S.Rept. 108-386. 116 p.

FOR ADDITIONAL READING

National Research Council. *Health Implications of Perchlorate Ingestion.* Committee to Assess the Health Implications of Perchlorate Ingestion. Board on Environmental Studies and Toxicology, Division of Earth and Life Studies. National Academy of Sciences. January 2005. 177 p.

U.S. Congress. Congressional Budget Office. *Future Investment in Drinking Water and Wastewater Infrastructure.* Nov. 2002. 56 p.

U.S. Environmental Protection Agency. *The Clean Water and Drinking Water Infrastructure Gap Analysis Report.* Report No. EPA 816-R-02-020. September 2002. 50 p.

In: Safe Drinking Water Act and its Interpretation ISBN: 1-59454-761-0
Editor: Thomas W. Carter, pp. 25-125 © 2006 Nova Science Publishers, Inc.

Chapter 2

1996 Amendments to the Safe Drinking Water Act: Public Law 104-182 104th Congress

Environmental Protection Agency

An Act

To reauthorize and amend title XIV of the Public Health Service Act (commonly known as the ``Safe Drinking Water Act''), and for other purposes. <<NOTE: Aug. 6, 1996 - [S. 1316]>>

Be it enacted by the Senate and House of Representatives of the United States of America in Congress <<NOTE: Safe Drinking Water Act Amendments of 1996. Inter-governmental relations. Environmental protection.>> assembled,

Section 1. Short Title; Table Of Contents

(a) <<NOTE: 42 USC 201 note.>> Short Title.--This Act may be cited as the ``Safe Drinking Water Act Amendments of 1996''.

Sec. 2. References; Effective Date; Disclaimer

(a) References to Safe Drinking Water Act.--Except as otherwise expressly provided, whenever in this Act an amendment or repeal is expressed in terms of an amendment to, or repeal of, a section or other provision, the reference shall be considered to be made to that section or other provision of title XIV of the Public Health Service Act (commonly known as the ``Safe Drinking Water Act'') (42 U.S.C. 300f et seq.).

(b) <<NOTE: 42 USC 300f note.>> Effective Date.--Except as otherwise specified in this Act or in the amendments made by this Act, this Act and the amendments made by this Act shall take effect on the date of enactment of this Act.

(c) <<NOTE: 42 USC 300f note.>> Disclaimer.--Except for the provisions of section 302 (relating to transfers of funds), nothing in this Act or in any amendments made by this Act to title XIV of the Public Health Service Act (commonly known as the ``Safe Drinking Water Act'') or any other law shall be construed by the Administrator of the Environmental Protection Agency or the courts as affecting, modifying, expanding, changing, or altering—

1. the provisions of the Federal Water Pollution Control Act;
2. the duties and responsibilities of the Administrator under that Act; or
3. the regulation or control of point or nonpoint sources of pollution discharged into waters covered by that Act.

The Administrator shall identify in the agency's annual budget all funding and full-time equivalents administering such title XIV separately from funding and staffing for the Federal Water Pollution Control Act.

Sec. 3. <<Note: 42 USC 300f note.>> Findings

The Congress finds that—

[1] safe drinking water is essential to the protection of public health;

[2] because the requirements of the Safe Drinking Water Act (42 U.S.C. 300f et seq.) now exceed the financial and technical capacity of some public water systems, especially many small public water systems, the Federal Government needs to provide assistance to

communities to help the communities meet Federal drinking water requirements;

[3] the Federal Government commits to maintaining and improving its partnership with the States in the administration and implementation of the Safe Drinking Water Act;

[4] States play a central role in the implementation of safe drinking water programs, and States need increased financial resources and appropriate flexibility to ensure the prompt and effective development and implementation of drinking water programs;

[5] the existing process for the assessment and selection of additional drinking water contaminants needs to be revised and improved to ensure that there is a sound scientific basis for setting priorities in establishing drinking water regulations;

[6] procedures for assessing the health effects of contaminants establishing drinking water standards should be revised to provide greater opportunity for public education and participation;

[7] in considering the appropriate level of regulation for contaminants in drinking water, risk assessment, based on sound and objective science, and benefit-cost analysis are important analytical tools for improving the efficiency and effectiveness of drinking water regulations to protect human health;

[8] more effective protection of public health requires--

(A) a Federal commitment to set priorities that will allow scarce Federal, State, and local resources to be targeted toward the drinking water problems of greatest public health concern;

(B) maximizing the value of the different and complementary strengths and responsibilities of the Federal and State governments in those States that have primary enforcement responsibility for the Safe Drinking Water Act; and

(C) prevention of drinking water contamination through well-trained system operators, water systems with adequate managerial, technical, and financial capacity, and enhanced protection of source waters of public water systems;

[9] compliance with the requirements of the Safe Drinking Water Act continues to be a concern at public water systems experiencing technical and financial limitations, and Federal, State, and local governments need more resources and more effective authority to attain the objectives of the Safe Drinking Water Act; and

[10] consumers served by public water systems should be provided with information on the source of the water they are drinking and its

quality and safety, as well as prompt notification of any violation of drinking water regulations.

TITLE I--AMENDMENTS TO SAFE DRINKING WATER ACT

Sec. 101. Definitions

(a) In General.--Section 1401 (42 U.S.C. 300f) is amended as follows:

(1) In paragraph (1)--

(A) in subparagraph (D), by inserting ``accepted methods for" before ``quality control"; and

(B) <<NOTE: Federal Register, publication.>> by adding at the end the following: ``At any time after promulgation of a regulation referred to in this paragraph, the Administrator may add equally effective quality control and testing procedures by guidance published in the Federal Register. Such procedures shall be treated as an alternative for public water systems to the quality control and testing procedures listed in the regulation.".

(2) In paragraph (13)--

(A) by striking ``The" and inserting ``(A) Except as provided in subparagraph (B), the"; and

(B) by adding at the end the following:

"(B) For purposes of section 1452, the term `State' means each of the 50 States, the District of Columbia, and the Commonwealth of Puerto Rico.".

(3) In paragraph (14), by adding at the end the following: ``For purposes of section 1452, the term includes any Native village (as defined in section 3(c) of the Alaska Native Claims Settlement Act (43 U.S.C. 1602(c))).".

(4) By adding at the end the following:

``(15) Community water system.--The term `community water system' means a public water system that--

"(A) serves at least 15 service connections used by year-round residents of the area served by the system; or

"(B) regularly serves at least 25 year-round residents.

``(16) Noncommunity water system.--The term `noncommunity water system' means a public water system that is not a community water system.".

(b) Public Water System.--

(1) In general.--Section 1401(4) (42 U.S.C. 300f(4)) is amended as follows:

(A) In the first sentence, by striking ``piped water for human consumption'' and inserting ``water for human consumption through pipes or other constructed conveyances''.

(B) By redesignating subparagraphs (A) and (B) as clauses (i) and (ii), respectively.

(C) By striking ``(4) The'' and inserting the following:

``(4) Public water system.--

``(A) In general.--The''; and

(D) by adding at the end the following:

``(B) Connections.--

``(i) In general.--For purposes of subparagraph (A), a connection to a system that delivers water by a constructed conveyance other than a pipe shall not be considered a connection, if--

``(I) the water is used exclusively for purposes other than residential uses (consisting of drinking, bathing, and cooking, or other similar uses);

``(II) the Administrator or the State (in the case of a State exercising primary enforcement responsibility for public water systems) determines that alternative water to achieve the equivalent level of public health protection provided by the applicable national primary drinking water regulation is provided for residential or similar uses for drinking and cooking; or

``(III) the Administrator or the State (in the case of a State exercising primary enforcement responsibility for public water systems) determines that the water provided for residential or similar uses for drinking, cooking, and bathing is centrally treated or treated at the point of entry by the provider, a pass-through entity, or the user to achieve the equivalent level of protection provided by the applicable national primary drinking water regulations.

``(ii) Irrigation districts.--An irrigation district in existence prior to May 18, 1994, that provides primarily agricultural service through a piped water system with only incidental residential or similar use shall not be considered to be a public water system if the system or the residential or similar users of the system comply with subclause (II) or (III) of clause (i).

``(C) Transition period.--A water supplier that would be a public water system only as a result of modifications made to this paragraph by the Safe Drinking Water Act Amendments of 1996 shall not be considered a public water system for purposes of the Act until the date that is two years after the date of enactment of this subparagraph. If a water supplier does not serve 15 service connections (as defined in subparagraphs (A) and (B)) or 25 people at any time after the conclusion of the 2-year period, the water supplier shall not be considered a public water system.''.

(2) <<NOTE: 42 USC 300f note.>> GAO study.--The Comptroller General of the United States shall undertake a study to--

(A) ascertain the numbers and locations of individuals and households relying for their residential water needs, including drinking, bathing, and cooking (or other similar uses) on irrigation water systems, mining water systems, industrial water systems, or other water systems covered by section 1401(4)(B) of the Safe Drinking Water Act that are not public water systems subject to the Safe Drinking Water Act;

(B) determine the sources and costs and affordability (to users and systems) of water used by such populations for their residential water needs; and

(C) review State and water system compliance with the exclusion provisions of section 1401(4)(B) of such Act.

The <<NOTE: Reports.>> Comptroller General shall submit a report to the Congress within 3 years after the date of enactment of this Act containing the results of such study.

Sec. 102. General Authority

(a) Standards.--Section 1412(b) (42 U.S.C. 300g-1(b)) is amended by striking ``(b)(1)'' and all that follows through the end of paragraph (3) and inserting the following:

``(b) Standards.--

``(1) Identification of contaminants for listing.--

``(A) <<NOTE: Publication. Regulations.>> General authority.--The Administrator shall, in accordance with the procedures established by this subsection, publish a maximum contaminant level goal and promulgate a national primary drinking water regulation for a contaminant (other than a contaminant referred to in paragraph (2) for which a national primary drinking water regulation has been promulgated as of the date of enactment of the Safe Drinking Water Act Amendments of 1996) if the Administrator determines that--

``(i) the contaminant may have an adverse effect on the health of persons;

``(ii) the contaminant is known to occur or there is a substantial likelihood that the contaminant will occur in public water systems with a frequency and at levels of public health concern; and

``(iii) in the sole judgment of the Administrator, regulation of such contaminant presents a meaningful opportunity for health risk reduction for persons served by public water systems.

``(B) Regulation of unregulated contaminants.--

``(i) <<NOTE: Publication.>> Listing of contaminants for consideration.--(I) Not later than 18 months after the date of enactment of the Safe Drinking Water Act Amendments of 1996 and every 5 years thereafter, the Administrator, after consultation with the scientific community, including the Science Advisory Board, after notice and opportunity for public comment, and after considering the occurrence data base established under section 1445(g), shall publish a list of contaminants which, at the time of publication, are not subject to any proposed or promulgated national primary drinking water regulation, which are known or anticipated to occur in public water systems, and which may require regulation under this title.

``(II) The unregulated contaminants considered under subclause (I) shall include, but not be limited to, substances referred to in section 101(14) of the Comprehensive Environmental Response, Compensation, and Liability Act of 1980, and substances registered as pesticides under the Federal Insecticide, Fungicide, and Rodenticide Act.

``(III) The Administrator's decision whether or not to select an unregulated contaminant for a list under this clause shall not be subject to judicial review.

``(ii) Determination to regulate.--(I) Not later than 5 years after the date of enactment of the Safe Drinking Water Act Amendments of 1996, and every 5 years thereafter, the Administrator shall, after notice of the preliminary determination and opportunity for public comment, for not fewer than 5 contaminants included on the list published under clause (i), make determinations of whether or not to regulate such contaminants.

``(II) A determination to regulate a contaminant shall be based on findings that the criteria of clauses (i), (ii), and (iii) of subparagraph (A) are satisfied. Such findings shall be based on the best available public health information, including the occurrence data base established under section 1445(g).

``(III) The Administrator may make a determination to regulate a contaminant that does not appear on a list under clause (i) if the determination to regulate is made pursuant to subclause (II).

``(IV) A determination under this clause not to regulate a contaminant shall be considered final agency action and subject to judicial review.

``(iii) Review.--Each document setting forth the determination for a contaminant under clause (ii) shall be available for public comment at such time as the determination is published.

``(C) Priorities.--In selecting unregulated contaminants for consideration under subparagraph (B), the Administrator shall select contaminants that present the greatest public health concern. The Administrator, in making such selection, shall take into consideration, among other factors of public health concern, the effect of such contaminants upon subgroups that comprise a meaningful portion of the general population (such as infants, children, pregnant women, the elderly, individuals with a history of serious illness, or other subpopulations) that are identifiable as being at greater risk of adverse health effects due to exposure to contaminants in drinking water than the general population.

``(D) Urgent threats to public health.--The Administrator may promulgate an interim national primary drinking water regulation for a contaminant without making a determination for the contaminant under paragraph (4)(C), or completing the analysis under paragraph (3)(C), to address an urgent threat to public health as determined by the Administrator after consultation with and written response to any comments provided by the Secretary of Health and Human Services, acting through the director of the Centers for Disease Control and Prevention or the director of the National Institutes of Health. <<NOTE: Publication.>> A determination for any contaminant in accordance with paragraph (4)(C) subject to an interim regulation under this subparagraph shall be issued, and a completed analysis meeting the requirements of paragraph (3)(C) shall be published, not later than 3 years after the date on which the regulation is promulgated and the regulation shall be repromulgated, or revised if appropriate, not later than 5 years after that date.

``(E) <<NOTE: Publication.>> Regulation.--For each contaminant that the Administrator determines to regulate under subparagraph (B), the Administrator shall publish maximum contaminant level goals and promulgate, by rule, national primary drinking water regulations under this subsection. The Administrator shall propose the maximum contaminant level goal and national primary drinking water regulation for a contaminant not later than 24 months after the determination to regulate under subparagraph (B), and may publish such proposed regulation concurrent with the determination to regulate. The Administrator shall publish a maximum contaminant level goal and promulgate a national primary drinking water regulation within 18 months after the proposal thereof. <<NOTE: Federal Register, publication.>> The Administrator, by notice in the Federal Register, may extend the deadline for such promulgation for up to 9 months.

``(F) Health advisories and other actions.--The Administrator may publish health advisories (which are not regulations) or take other appropriate actions for contaminants not subject to any national primary drinking water regulation.

``(2) Schedules and deadlines.--

``(A) <<NOTE: Publication. Regulations.>> In general.--In the case of the contaminants listed in the Advance Notice of Proposed Rulemaking published in volume 47, Federal Register, page 9352, and in volume 48, Federal Register, page 45502, the Administrator shall publish maximum contaminant level goals and promulgate national primary drinking water regulations--

``(i) not later than 1 year after June 19, 1986, for not fewer than 9 of the listed contaminants;

``(ii) not later than 2 years after June 19, 1986, for not fewer than 40 of the listed contaminants; and

``(iii) not later than 3 years after June 19, 1986, for the remainder of the listed contaminants.

``(B) Substitution of contaminants.--If the Administrator identifies a drinking water contaminant the regulation of which, in the judgment of the Administrator, is more likely to be protective of public health (taking into account the schedule for regulation under subparagraph (A)) than a contaminant referred to in subparagraph (A), the Administrator may publish a maximum contaminant level goal and promulgate a national primary drinking water regulation for the identified contaminant in lieu of regulating the contaminant referred to in subparagraph (A). Substitutions may be made for not more than 7 contaminants referred to in subparagraph (A). Regulation of a contaminant identified under this subparagraph shall be in accordance with the schedule applicable to the contaminant for which the substitution is made.

``(C) <<NOTE: Rules.>> Disinfectants and disinfection byproducts.--The Administrator shall promulgate an Interim Enhanced Surface Water Treatment Rule, a Final Enhanced Surface Water Treatment

Rule, a Stage I Disinfectants and Disinfection Byproducts Rule, and a Stage II Disinfectants and Disinfection Byproducts Rule in accordance with the schedule published in volume 59, Federal Register, page 6361 (February 10, 1994), in table III.13 of the proposed Information Collection Rule. If a delay occurs with respect to the promulgation of any rule in the schedule referred to in this subparagraph, all subsequent rules shall be completed as expeditiously as practicable but no later than a revised date that reflects the interval or intervals for the rules in the schedule.".

(b) <<NOTE: 42 USC 300g-1 note.>> Applicability of Prior Requirements.--The requirements of subparagraphs (C) and (D) of section 1412(b)(3) of the Safe Drinking Water Act as in effect before the date of enactment of this Act, and any obligation to promulgate regulations pursuant to such subparagraphs not promulgated as of the date of enactment of this Act, are superseded by the amendments made by subsection (a).

(c) Conforming Amendments.--(1) Section 1415(d) (42 U.S.C. 300g-
4(d)) is amended by striking ``1412(b)(3)" and inserting ``1412(b)".

(2) Section 1412(a)(3) (42 U.S.C. 300g-1(a)(3)) is amended by striking ``paragraph (1), (2), or (3) of" in each place it appears.

Sec. 103. Risk Assessment, Management, and Communication

Section 1412(b) (42 U.S.C. 300g-1(b)) is amended by inserting after paragraph (2) the following:

``(3) Risk assessment, management, and communication.--

``(A) Use of science in decisionmaking.--In carrying out this section, and, to the degree that an Agency action is based on science, the Administrator shall use--

``(i) the best available, peer-reviewed science and supporting studies conducted in accordance with sound and objective scientific practices; and

``(ii) data collected by accepted methods or best available methods (if the reliability of the method and the nature of the decision justifies use of the data).

``(B) Public information.--In carrying out this section, the Administrator shall ensure that the presentation of information on public health effects is comprehensive, informative, and understandable. The Administrator shall, in a document made available to the public in support of a regulation promulgated under this section, specify, to the extent practicable--

``(i) each population addressed by any estimate of public health effects;

``(ii) the expected risk or central estimate of risk for the specific populations;

``(iii) each appropriate upper-bound or lower-bound estimate of risk;

``(iv) each significant uncertainty identified in the process of the assessment of public health effects and studies that would assist in resolving the uncertainty; and

``(v) peer-reviewed studies known to the Administrator that support, are directly relevant to, or fail to support any estimate of public health effects and the methodology used to reconcile inconsistencies in the scientific data.

``(C) Health risk reduction and cost analysis.--

``(i) <<NOTE: Publication.>> Maximum contaminant levels.--When proposing any national primary drinking water regulation that includes a maximum contaminant level, the Administrator shall, with respect to a maximum contaminant level that is being considered in accordance with paragraph (4) and each alternative maximum contaminant level that is being considered pursuant to paragraph (5) or (6)(A), publish, seek public comment on, and use for the purposes of paragraphs (4), (5), and (6) an analysis of each of the following:

``(I) Quantifiable and nonquantifiable health risk reduction benefits for which there is a factual basis in the rulemaking record to conclude that such benefits are likely to occur as the result of treatment to comply with each level.

``(II) Quantifiable and nonquantifiable health risk reduction benefits for which there is a factual basis in the rulemaking record to conclude that such benefits are likely to occur from reductions in co-occurring contaminants that may be attributed solely to compliance with the maximum contaminant level, excluding benefits resulting from compliance with other proposed or promulgated regulations.

``(III) Quantifiable and nonquantifiable costs for which there is a factual basis in the rulemaking record to conclude that such costs are likely to occur solely as a result of compliance with the maximum contaminant level, including monitoring, treatment, and other costs and excluding costs resulting from compliance with other proposed or promulgated regulations.

``(IV) The incremental costs and benefits associated with each alternative maximum contaminant level considered.

``(V) The effects of the contaminant on the general population and on groups within the general population such as infants, children, pregnant women, the elderly, individuals with a history of serious illness, or other subpopulations that are identified as likely to be at greater risk of adverse health effects due to exposure to contaminants in drinking water than the general population.

``(VI) Any increased health risk that may occur as the result of compliance, including risks associated with co-occurring contaminants.

``(VII) Other relevant factors, including the quality and extent of the information, the uncertainties in the analysis supporting subclauses (I) through (VI), and factors with respect to the degree and nature of the risk.

``(ii) <<NOTE: Publication.>> Treatment techniques.--When proposing a national primary drinking water regulation that includes a treatment technique in accordance with paragraph (7)(A), the Administrator shall publish and seek public comment on an analysis of the health risk reduction benefits and costs likely to be experienced as the result of compliance with the treatment technique and alternative treatment techniques that are being considered, taking into account, as appropriate, the factors described in clause (i).

``(iii) Approaches to measure and value benefits.--The Administrator may identify valid approaches for the measurement and valuation of benefits under this subparagraph, including approaches to identify consumer willingness to pay for reductions in health risks from drinking water contaminants.

``(iv) Authorization.--There are authorized to be appropriated to the Administrator, acting through the Office of Ground Water and Drinking Water, to conduct studies, assessments, and analyses in support of regulations or the development of methods, $35,000,000 for each of fiscal years 1996 through 2003.''.

Sec. 104. Standard-Setting

(a) In General.--Section 1412(b) (42 U.S.C. 300g-1(b)) is amended as follows:

(1) In paragraph (4)--

(A) by striking ``(4) Each" and inserting the following:

``(4) Goals and standards.--

``(A) Maximum contaminant level goals.--Each";

(B) in the last sentence--

(i) by striking ``Each national" and inserting the following:

``(B) Maximum contaminant levels.-- Except as provided in paragraphs (5) and (6), each national"; and

(ii) by striking ``maximum level" and inserting ``maximum contaminant level"; and

(C) by adding at the end the following:

``(C) <<NOTE: Publication.>> Determination.--At the time the Administrator proposes a national primary drinking water regulation under this paragraph, the Administrator shall publish a determination as to whether the benefits of the maximum contaminant level justify, or do not justify, the costs based on the analysis conducted under paragraph (3)(C).".

(2) By striking ``(5) For the" and inserting the following:

``(D) Definition of feasible.--For the".

(3) In the second sentence of paragraph (4)(D) (as so designated), by striking ``paragraph (4)" and inserting ``this paragraph".

(4) By striking ``(6) Each national" and inserting the following:

``(E) Feasible technologies.--

``(i) In general.--Each national".

(5) In paragraph (4)(E)(i) (as so designated), by striking ``this paragraph" and inserting ``this subsection".

(6) By inserting after paragraph (4) (as so amended) the following:

``(5) Additional health risk considerations.--

``(A) In general.--Notwithstanding paragraph (4), the Administrator may establish a maximum contaminant level for a contaminant at a level other than the feasible level, if the technology, treatment techniques, and other means used to determine the feasible level would result in an increase in the health risk from drinking water by--

``(i) increasing the concentration of other contaminants in drinking water; or

``(ii) interfering with the efficacy of drinking water treatment techniques or processes that are used to comply with other national primary drinking water regulations.

``(B) Establishment of level.--If the Administrator establishes a maximum contaminant level or levels or requires the use of treatment techniques for any contaminant or contaminants pursuant to the authority of this paragraph--

``(i) the level or levels or treatment techniques shall minimize the overall risk of adverse health effects by balancing the risk from the contaminant and the risk from other contaminants the concentrations of which may be affected by the use of a treatment technique or process that would be employed to attain the maximum contaminant level or levels; and

``(ii) the combination of technology, treatment techniques, or other means required to meet the level or levels shall not be more stringent than is feasible (as defined in paragraph (4)(D)).

``(6) Additional health risk reduction and cost considerations.--

``(A) In general.--Notwithstanding paragraph (4), if the Administrator determines based on an analysis conducted under paragraph (3)(C) that the benefits of a maximum contaminant level promulgated in accordance with paragraph (4) would not justify the costs of complying with the level, the Administrator may, after notice and opportunity for public comment, promulgate a maximum contaminant level for the contaminant that maximizes health risk reduction benefits at a cost that is justified by the benefits.

``(B) Exception.--The Administrator shall not use the authority of this paragraph to promulgate a maximum contaminant level for a contaminant, if the benefits of compliance with a national primary drinking water regulation for the contaminant that would be promulgated in accordance with paragraph (4) experienced by--

``(i) persons served by large public water systems; and

``(ii) persons served by such other systems as are unlikely, based on information provided by the States, to receive a variance under section 1415(e) (relating to small system variances);

would justify the costs to the systems of complying with the regulation. This subparagraph shall not apply if the contaminant is found almost exclusively in small systems eligible under section 1415(e) for a small system variance.

``(C) Disinfectants and disinfection byproducts.--

The Administrator may not use the authority of this paragraph to establish a maximum contaminant level in a Stage I or Stage II national primary drinking water regulation (as described in paragraph (2)(C)) for contaminants that are disinfectants or disinfection byproducts, or to establish a maximum contaminant level or treatment technique requirement for the control of cryptosporidium. The authority of this paragraph may be used to establish regulations for the use of disinfection by systems relying on ground water sources as required by paragraph (8).

``(D) Judicial review.--A determination by the Administrator that the benefits of a maximum contaminant level or treatment requirement justify or do not justify the costs of complying with the level shall be reviewed by the court pursuant to section 1448 only as part of a review of a final national primary drinking water regulation that has been promulgated based on the determination and shall not be set aside by the court under that section unless the court finds that the determination is arbitrary and capricious.''.

(b) <<NOTE: 42 USC 300g-1 note.>> Disinfectants and Disinfection Byproducts.--The Administrator of the Environmental Protection Agency may use the authority of section 1412(b)(5) of the Safe Drinking Water Act (as amended by this Act) to promulgate the Stage I and Stage II Disinfectants and Disinfection Byproducts Rules as proposed in volume 59, Federal Register, page 38668 (July 29, 1994). The considerations used in the development of the July 29, 1994, proposed national primary drinking water regulation on disinfectants and disinfection byproducts shall be treated as consistent with such section 1412(b)(5) for purposes of such Stage I and Stage II rules.

(c) Review of Standards.--Section 1412(b)(9) (42 U.S.C. 300g-1(b)(9)) is amended to read as follows:

``(9) Review and revision.--The Administrator shall, not less often than every 6 years, review and revise, as appropriate, each national primary drinking water regulation promulgated under this title. Any revision of a national primary drinking water regulation shall be promulgated in accordance with this section, except that each revision shall maintain, or provide for greater, protection of the health of persons.''.

Sec. 105. Treatment Technologies for Small Systems

Section 1412(b)(4)(E) (42 U.S.C. 300g-1(b)(4)(E)) is amended by adding at the end the following:

``(ii) List of technologies for small systems.--The Administrator shall include in the list any technology, treatment technique, or other means that is affordable, as determined by the Administrator in consultation with the States, for small public water systems serving--

``(I) a population of 10,000 or fewer but more than 3,300;

``(II) a population of 3,300 or fewer but more than 500; and

``(III) a population of 500 or fewer but more than 25;

and that achieves compliance with the maximum contaminant level or treatment technique, including packaged or modular systems and point-of-entry or point-of-use treatment units. Point-of-entry and point-of-use treatment units shall be owned, controlled and maintained by the public water system or by a person under contract with the public water system to ensure proper operation and maintenance and compliance with the maximum contaminant level or treatment technique and equipped with mechanical warnings to ensure that customers are automatically notified of operational problems. The Administrator shall not include in the list any point-of-use treatment technology, treatment technique, or other means to achieve compliance with a maximum contaminant level or treatment technique requirement for a microbial contaminant (or an indicator of a microbial contaminant). If the American National Standards Institute has issued product standards applicable to a specific type of point-of-entry or point-of-use treatment unit, individual units of that type shall not be accepted for compliance with a maximum contaminant level or treatment technique requirement unless they are independently certified in accordance with such standards. In listing any technology, treatment technique, or other means pursuant to this clause, the Administrator shall consider the quality of the source water to be treated.

``(iii) List of technologies that achieve compliance.--Except as provided in clause (v), not later than 2 years after the date of enactment of this clause and after consultation with the States, the Administrator shall issue a list of technologies that achieve compliance with the maximum contaminant level or treatment technique for each category of public water systems described in subclauses (I), (II), and (III) of clause (ii) for each national primary drinking water regulation promulgated prior to the date of enactment of this paragraph.

``(iv) Additional technologies.--The Administrator may, at any time after a national primary drinking water regulation has been promulgated, supplement the list of technologies describing additional or new or innovative treatment technologies that meet the requirements of this paragraph for categories of small public water systems described in subclauses (I), (II), and (III) of clause (ii) that are subject to the regulation.

``(v) <<NOTE: Records.>> Technologies that meet surface water treatment rule.--Within one year after the date of enactment of this clause, the Administrator shall list technologies that meet the Surface Water Treatment Rule for each category of public water systems described in subclauses (I), (II), and (III) of clause (ii).''.

Sec. 106. Limited Alternative to Filtration

Section 1412(b)(7)(C) (42 U.S.C. 300g-1(b)(7)(C)) is amended by adding the following after clause (iv):

``(v) As an additional alternative to the regulations promulgated pursuant to clauses (i) and (iii), including the criteria for avoiding filtration contained in 40 CFR 141.71, a State exercising primary enforcement responsibility for public water systems may, on a case-by-case basis, and after notice and opportunity for public comment, establish treatment requirements as an alternative to filtration in the case of systems having uninhabited, undeveloped watersheds in consolidated ownership, and having control over access to, and activities in, those watersheds, if the State determines (and the Administrator concurs) that the quality of the source water and the alternative treatment requirements established by the State ensure greater removal or inactivation efficiencies of pathogenic organisms for which national primary drinking water regulations have been promulgated or that are of public health concern than would be achieved by the combination of filtration and chlorine disinfection (in compliance with this section).''.

Sec. 107. <<Note: Regulations.>> Ground Water Disinfection

Paragraph (8) of section 1412(b) (42 U.S.C. 300g-1(b)(8)) is amended by moving the margins of such paragraph 2 ems to the right and by striking the first sentence and inserting the following: ``Disinfection.--At any time after the end of the 3-year period that begins on the date of enactment of the Safe Drinking Water Act Amendments of 1996, but not later than the date on which the Administrator promulgates a Stage II rulemaking for disinfectants and disinfection byproducts (as described in paragraph (2)(C)), the Administrator shall also promulgate national primary drinking water regulations requiring disinfection as a treatment technique for all public water systems, including surface water systems and, as necessary, ground water systems. After consultation with the States, the Administrator shall (as part of the regulations) promulgate criteria that the Administrator, or a State that has primary enforcement responsibility under section 1413, shall apply to determine whether disinfection shall be required as a treatment technique for any public water system served by ground water.''.

Sec. 108. Effective Date for Regulations

Section 1412(b)(10) (42 U.S.C. 300g-1(b)(10)) is amended to read as follows:

``(10) Effective date.--A national primary drinking water regulation promulgated under this section (and any amendment thereto) shall take effect on the date that is 3 years after the date on which the regulation is promulgated unless the Administrator determines that an earlier date is practicable, except that the Administrator, or a State (in the case of an individual system), may allow up to 2 additional years to comply with a maximum contaminant level or treatment technique if the Administrator or State (in the case of an individual system) determines that additional time is necessary for capital improvements.''.

Sec. 109. Arsenic, Sulfate, and Radon

(a) Arsenic and Sulfate.--Section 1412(b) (42 U.S.C. 300g-1(b)) is amended by inserting after paragraph (11) the following:

``(12) Certain contaminants.--

``(A) Arsenic.--

``(i) Schedule and standard.--Notwithstanding the deadlines set forth in paragraph (1), the Administrator shall promulgate a national primary drinking water regulation for arsenic pursuant to this subsection, in accordance with the schedule established by this paragraph.

``(ii) Study plan.--Not later than 180 days after the date of enactment of this paragraph, the Administrator shall develop a comprehensive plan for study in support of drinking water rulemaking to reduce the uncertainty in assessing health risks associated with exposure to low levels of arsenic. In conducting such study, the Administrator shall consult with the National Academy of Sciences, other Federal agencies, and interested public and private entities.

``(iii) Cooperative agreements.--In carrying out the study plan, the Administrator may enter into cooperative agreements with other Federal agencies, State and local governments, and other interested public and private entities.

``(iv) Proposed regulations.--The Administrator shall propose a national primary drinking water regulation for arsenic not later than January 1, 2000.

``(v) Final regulations.--Not later than January 1, 2001, after notice and opportunity for public comment, the Administrator shall promulgate a national primary drinking water regulation for arsenic.

``(vi) Authorization.--There are authorized to be appropriated $2,500,000 for each of fiscal years 1997 through 2000 for the studies required by this paragraph.

``(B) Sulfate.--

``(i) Additional study.--Prior to promulgating a national primary drinking water regulation for sulfate, the Administrator and the Director of the Centers for Disease Control and Prevention shall jointly conduct an additional study to establish a reliable dose-response relationship for the adverse human health effects that may result from exposure to sulfate in drinking water, including the health effects that may be experienced by groups within the general population (including infants and travelers) that are potentially at greater risk of adverse health effects as the result of such exposure. The study shall be conducted in consultation with interested States, shall be based on the best available, peer-reviewed science and supporting studies conducted in accordance with sound and objective scientific practices, and shall be completed not later than 30 months after the date of enactment of the Safe Drinking Water Act Amendments of 1996.

``(ii) Determination.--The Administrator shall include sulfate among the 5 or more contaminants for which a determination is made pursuant to paragraph (3)(B) not later than 5 years after the date of enactment of the Safe Drinking Water Act Amendments of 1996.

``(iii) Proposed and final rule.--Notwithstanding the deadlines set forth in paragraph (2), the Administrator may, pursuant to the authorities of this subsection and after notice and opportunity for public comment, promulgate a final national primary drinking water regulation for sulfate. Any such regulation shall include requirements for public notification and options for the provision of alternative water supplies to populations at risk as a means of complying with the regulation in lieu of a best available treatment technology or other means.''.

(b) Radon.--Section 1412(b) (42 U.S.C. 300g-1(b)) is amended by inserting after paragraph (12) the following:

``(13) Radon in drinking water.--

``(A) National primary drinking water regulation.--Notwithstanding paragraph (2), the Administrator shall withdraw any national primary drinking water regulation for radon proposed prior to the date of enactment of this paragraph and shall propose and promulgate a regulation for radon under this section, as amended by the Safe Drinking Water Act Amendments of 1996.

``(B) Risk assessment and studies.--

``(i) Assessment by nas.--Prior to proposing a national primary drinking water regulation for radon, the Administrator shall arrange for the National Academy of Sciences to prepare a risk assessment for radon in drinking water using the best available science in accordance with the requirements of paragraph (3). The risk assessment shall consider each of the risks associated with exposure to radon from drinking water and consider studies on the health effects of radon at levels and under conditions likely to be experienced through residential exposure. The risk assessment shall be peer-reviewed.

``(ii) Study of other measures.--The Administrator shall arrange for the National Academy of Sciences to prepare an assessment of the health risk reduction benefits associated with various mitigation measures to reduce radon levels in indoor air. The assessment may be conducted as part of the risk assessment authorized by clause (i) and shall be used by the Administrator to prepare the guidance and approve State programs under subparagraph (G).

``(iii) <<NOTE: Contracts.>> Other organization.--If the National Academy of Sciences declines to prepare the risk assessment or studies required by this subparagraph, the Administrator shall enter into a contract or cooperative agreement with another independent, scientific organization to prepare such assessments or studies.

``(C) <<NOTE: Publication.>> Health risk reduction and cost analysis.--Not later than 30 months after the date of enactment of this paragraph, the Administrator shall publish, and seek public comment on, a health risk reduction and cost analysis meeting the requirements of paragraph (3)(C) for potential maximum contaminant levels that are being considered for radon in drinking water. The Administrator shall include a response to all significant public comments received on the analysis with the preamble for the proposed rule published under subparagraph (D).

``(D) Proposed regulation.--Not later than 36 months after the date of enactment of this paragraph, the Administrator shall propose a maximum contaminant level goal and a national primary drinking water regulation for radon pursuant to this section.

``(E) Final regulation.--Not later than 12 months after the date of the proposal under subparagraph (D), the Administrator shall publish a maximum contaminant level goal and promulgate a national primary drinking water regulation for radon pursuant to this section based on the risk assessment prepared pursuant to subparagraph (B) and the health risk reduction and cost analysis published pursuant to subparagraph (C). In considering the risk assessment and the health risk reduction and cost analysis in connection with the promulgation of such a standard, the Administrator shall take into account the costs and benefits of control programs for radon from other sources.

``(F) <<NOTE: Regulations.>> Alternative maximum contaminant level.--If the maximum contaminant level for radon in drinking water promulgated pursuant to subparagraph (E) is more stringent than necessary to reduce the contribution to radon in indoor air from drinking water to a concentration that is equivalent to the national average concentration of radon in outdoor air, the Administrator shall, simultaneously with the promulgation of such level, promulgate an alternative maximum contaminant level for radon that would result in a contribution of radon from drinking water to radon levels in indoor air equivalent to the national average concentration of radon in outdoor air. <<NOTE: Publication. Guidelines.>> If the Administrator promulgates an alternative maximum contaminant level under this subparagraph, the Administrator shall, after notice and opportunity for public comment and in consultation with the States, publish guidelines for State programs, including criteria for multimedia measures to mitigate radon levels in indoor air, to be used by the States in preparing programs under subparagraph (G). The guidelines shall take into account data from existing radon mitigation programs and the assessment of mitigation measures prepared under subparagraph (B).

``(G) Multimedia radon mitigation programs.--

``(i) In general.--A State may develop and submit a multimedia program to mitigate radon levels in indoor air for approval by the Administrator under this subparagraph. If, after notice and the opportunity for public comment, such program is approved by the Administrator, public water systems in the State may comply with the alternative maximum contaminant level promulgated under subparagraph (F) in lieu of the maximum contaminant level in the national primary drinking water regulation promulgated under subparagraph (E).

``(ii) Elements of programs.--State programs may rely on a variety of mitigation measures including public education, testing, training, technical assistance, remediation grant and loan or incentive programs, or other regulatory or nonregulatory measures. The effectiveness of elements in State programs shall be evaluated by the Administrator based on the assessment prepared by the National Academy of Sciences under subparagraph (B) and the guidelines published by the Administrator under subparagraph (F).

``(iii) Approval.--The Administrator shall approve a State program submitted under this paragraph if the health risk reduction benefits expected to be achieved by the program are equal to or greater than the health risk reduction benefits that would be achieved if each public water system in the State complied with the maximum contaminant level promulgated under subparagraph (E). The Administrator shall approve or disapprove a program submitted under this paragraph within 180 days of receipt. A program that is not disapproved during such period shall be deemed approved. A program that is disapproved may be modified to address the objections of the Administrator and be resubmitted for approval.

``(iv) Review.--The Administrator shall periodically, but not less often than every 5 years, review each multimedia mitigation program approved under this subparagraph to determine whether it continues to meet the requirements of clause (iii) and shall, after written notice to the State and an opportunity for the State to correct any deficiency in the program, withdraw approval of programs that no longer comply with such requirements.

``(v) Extension.--If, within 90 days after the promulgation of an alternative maximum contaminant level under subparagraph (F), the Governor of a State submits a letter to the Administrator committing to develop a multimedia mitigation program under this subparagraph, the effective date of the national primary drinking water regulation for radon in the State that would be applicable under paragraph (10) shall be extended for a period of 18 months.

``(vi) Local programs.--In the event that a State chooses not to submit a multimedia mitigation program for approval under this subparagraph or has submitted a program that has been disapproved, any public water system in the State may submit a program for approval by the Administrator according to the same criteria, conditions, and approval process that would apply to a State program. The Administrator shall approve a multimedia mitigation program if the health risk reduction benefits expected to be achieved by the program are equal to or greater than the health risk reduction benefits that would result from compliance by the public water system with the maximum contaminant level for radon promulgated under subparagraph (E).".

Sec. 110. <<Note: Regulations.>> Recycling of Filter Backwash

Section 1412(b) (42 U.S.C. 300g-1(b)) is amended by adding the following new paragraph after paragraph (13):

``(14) Recycling of filter backwash.--The Administrator shall promulgate a regulation to govern the recycling of filter backwash water within the treatment process of a public water system. The Administrator shall promulgate such regulation not later than 4 years after the date of enactment of the Safe Drinking Water Act Amendments of 1996 unless such recycling has been addressed by the Administrator's Enhanced Surface Water Treatment Rule prior to such date.".

Sec. 111. Technology and Treatment Techniques

(a) Variance Technologies.--Section 1412(b) (42 U.S.C. 300g-1(b)) is amended by adding the following new paragraph after paragraph (14):

``(15) <<NOTE: Regulations.>> Variance technologies.--

``(A) In general.--At the same time as the Administrator promulgates a national primary drinking water regulation for a contaminant pursuant to this section, the Administrator shall issue guidance or regulations describing the best treatment technologies, treatment techniques, or other means (referred to in this paragraph as `variance technology') for the contaminant that the Administrator finds, after examination for efficacy under field conditions and not solely under laboratory conditions, are available and affordable, as determined by the Administrator in consultation with the States, for public water systems of varying size, considering the quality of the source water to be treated. The Administrator shall identify such variance technologies for public water systems serving--

``(i) a population of 10,000 or fewer but more than 3,300;

``(ii) a population of 3,300 or fewer but more than 500; and

``(iii) a population of 500 or fewer but more than 25,

if, considering the quality of the source water to be treated, no treatment technology is listed for public water systems of that size under paragraph (4)(E). Variance technologies identified by the Administrator pursuant to this paragraph may not achieve compliance with the maximum contaminant level or treatment technique requirement of such regulation, but shall achieve the maximum reduction or inactivation efficiency that is affordable considering the size of the system and the quality of the source water. The guidance or regulations shall not require the use of a technology from a specific manufacturer or brand.

``(B) Limitation.--The Administrator shall not identify any variance technology under this paragraph, unless the Administrator has determined, considering the quality of the source water to be treated and the expected useful life of the technology, that the variance technology is protective of public health.

``(C) Additional information.--The Administrator shall include in the guidance or regulations identifying variance technologies under this paragraph any assumptions supporting the public health determination referred to in subparagraph (B), where such assumptions concern the public water system to which the technology may be applied, or its source waters. The Administrator shall provide any assumptions used in determining affordability, taking into consideration the number of persons served by such systems. The Administrator shall provide as much reliable information as practicable on performance, effectiveness, limitations, costs, and other relevant factors including the applicability of variance technology to waters from surface and underground sources.

``(D) Regulations and guidance.--Not later than 2 years after the date of enactment of this paragraph and after consultation with the States, the Administrator shall issue guidance or regulations under subparagraph (A) for each national primary drinking water regulation promulgated prior to the date of enactment of this paragraph for which a variance may be granted under section 1415(e). The Administrator may, at any time after a national primary drinking water regulation has been promulgated, issue guidance or regulations describing additional variance technologies. The Administrator shall, not less often than every 7 years, or upon receipt of a petition supported by substantial information, review variance technologies identified under this paragraph. The Administrator shall issue revised guidance or regulations if new or innovative variance technologies become available that meet the requirements of this paragraph and achieve an equal or greater reduction or inactivation efficiency than the variance technologies previously identified under this subparagraph. No public water system shall be required to replace a variance technology during the useful life of the technology for the sole reason that a more efficient variance technology has been listed under this subparagraph.''.

(b) Availability of Information on Small System Technologies.--

Section 1445 (42 U.S.C. 300j-4) is amended by adding the following new subsection after subsection (g):

``(h) Availability of Information on Small System Technologies.-- For purposes of sections 1412(b)(4)(E) and 1415(e) (relating to small system variance program), the Administrator may request information on the characteristics of commercially available treatment systems and technologies, including the effectiveness and performance of the systems and technologies under various operating conditions. The Administrator may specify the form, content, and submission date of information to be submitted by manufacturers, States, and other interested persons for the purpose of considering the systems and technologies in the development of regulations or guidance under sections 1412(b)(4)(E) and 1415(e).''.

Sec. 112. State Primacy

(a) State Primary Enforcement Responsibility.--Section 1413 (42 U.S.C. 300g-2) is amended as follows:

(1) In subsection (a), by amending paragraph (1) to read as follows:

``(1) has adopted drinking water regulations that are no less stringent than the national primary drinking water regulations promulgated by the Administrator under subsections (a) and (b) of section 1412 not later than 2 years after the date on which the regulations are promulgated by the Administrator, except that the Administrator may provide for an extension of not more than 2 years if, after submission and review of appropriate, adequate documentation from the State, the Administrator determines that the extension is necessary and justified;''.

(2) By adding at the end the following subsection:

``(c) Interim Primary Enforcement Authority.--A State that has primary enforcement authority under this section with respect to each existing national primary drinking water regulation shall be considered to have primary enforcement authority with respect to each new or revised national primary drinking water regulation during the period beginning on the effective date of a regulation adopted and submitted by the State with respect to the new or revised national primary drinking water regulation in accordance with subsection (b)(1) and ending at such time as the Administrator makes a determination under subsection (b)(2)(B) with respect to the regulation.''.

(b) Emergency Plans.--Section 1413(a)(5) (42 U.S.C. 300g-2(a)(5)) is amended by inserting after ``emergency circumstances''

the following: ``including earthquakes, floods, hurricanes, and other natural disasters, as appropriate''.

Sec. 113. Enforcement; Judicial Review

(a) In General.--Section 1414 (42 U.S.C. 300g-3) is amended as follows:

(1) In subsection (a)--

(A) in paragraph (1)--

(i) in subparagraph (A)--

(I) in clause (i), by striking ``any national primary drinking water regulation in effect under section 1412'' and inserting ``any applicable requirement''; and

(II) by striking ``with such regulation or requirement'' and inserting ``with the requirement''; and

(ii) in subparagraph (B), by striking ``regulation or'' and inserting ``applicable''; and

(B) by striking paragraph (2) and inserting the following:

``(2) Enforcement in nonprimacy states.--

``(A) In general.--If, on the basis of information available to the Administrator, the Administrator finds, with respect to a period in which a State does not have primary enforcement responsibility for public water systems, that a public water system in the State--

``(i) for which a variance under section 1415 or an exemption under section 1416 is not in effect, does not comply with any applicable requirement; or

``(ii) for which a variance under section 1415 or an exemption under section 1416 is in effect, does not comply with any schedule or other requirement imposed pursuant to the variance or exemption;

the <<NOTE: Orders.>> Administrator shall issue an order under subsection (g) requiring the public water system to comply with the requirement, or commence a civil action under subsection (b).

``(B) Notice.--If the Administrator takes any action pursuant to this paragraph, the Administrator shall notify an appropriate local elected official, if any, with jurisdiction over the public water system of the action prior to the time that the action is taken.''.

(2) In the first sentence of subsection (b), by striking ``a national primary drinking water regulation'' and inserting ``any applicable requirement''.

(3) In subsection (g)--

(A) in paragraph (1), by striking ``regulation, schedule, or other'' each place it appears and inserting ``applicable'';

(B) in paragraph (2)--

(i) in the first sentence--

(I) by striking ``effect until after notice and opportunity for public hearing and,'' and inserting ``effect,''; and

(II) by striking ``proposed order'' and inserting ``order''; and

(ii) in the second sentence, by striking ``proposed to be''; and

(C) in paragraph (3)--

(i) by striking subparagraph (B) and inserting the following:

``(B) In a case in which a civil penalty sought by the Administrator under this paragraph does not exceed $5,000, the penalty shall be assessed by the Administrator after notice and opportunity for a public hearing (unless the person against whom the penalty is assessed requests a hearing on the record in accordance with section 554 of title 5, United States Code). In a case in which a civil penalty sought by the Administrator under this paragraph exceeds $5,000, but does not exceed $25,000, the penalty shall be assessed by the Administrator after notice and opportunity for a hearing on the record in accordance with section 554 of title 5, United States Code.''; and

(ii) in subparagraph (C), by striking ``paragraph exceeds $5,000'' and inserting ``subsection for a violation of an applicable requirement exceeds $25,000''.

(4) By adding at the end the following:

``(h) Consolidation Incentive.--

``(1) In general.--An owner or operator of a public water system may submit to the State in which the system is located (if the State has primary enforcement responsibility under section 1413) or to the Administrator (if the State does not have primary enforcement responsibility) a plan (including specific measures and schedules) for

``(A) the physical consolidation of the system with 1 or more other systems;

``(B) the consolidation of significant management and administrative functions of the system with 1 or more other systems; or

``(C) the transfer of ownership of the system that may reasonably be expected to improve drinking water quality.

``(2) Consequences of approval.--If the State or the Administrator approves a plan pursuant to paragraph (1), no enforcement action shall be taken pursuant to this part with respect to a specific violation identified in the approved plan prior to the date that is the earlier of the date on which consolidation is completed according to the plan or the date that is 2 years after the plan is approved.

``(i) Definition of Applicable Requirement.--In this section, the term `applicable requirement' means--

``(1) a requirement of section 1412, 1414, 1415, 1416, 1417, 1441, or 1445;

``(2) a regulation promulgated pursuant to a section referred to in paragraph (1);

``(3) a schedule or requirement imposed pursuant to a section referred to in paragraph (1); and

``(4) a requirement of, or permit issued under, an applicable State program for which the Administrator has made a determination that the requirements of section 1413 have been satisfied, or an applicable State program approved pursuant to this part.''.

(b) State Authority for Administrative Penalties.--Section 1413(a) (42 U.S.C. 300g-2(a)) is amended--

(1) by striking ``and'' at the end of paragraph (4);

(2) by striking the period at the end of paragraph (5) and inserting ``; and''; and

(3) by adding at the end the following:

``(6) has adopted authority for administrative penalties (unless the constitution of the State prohibits the adoption of the authority) in a maximum amount--

``(A) in the case of a system serving a population of more than 10,000, that is not less than $1,000 per day per violation; and

``(B) in the case of any other system, that is adequate to ensure compliance (as determined by the State);

except that a State may establish a maximum limitation on the total amount of administrative penalties that may be imposed on a public water system per violation.''.

(c) Judicial Review.--Section 1448(a) (42 U.S.C. 300j-7(a)) is amended--

(1) in paragraph (2) of the first sentence, by inserting ``final'' after ``any other'';

(2) in the second sentence, by striking ``or issuance of the order'' and inserting ``or any other final Agency action''; and

(3) by adding at the end the following ``In any petition concerning the assessment of a civil penalty pursuant to section 1414(g)(3)(B), the petitioner shall simultaneously send a copy of the complaint by certified mail to the Administrator and the Attorney General. The court shall set aside and remand the penalty order if the court finds that there is not substantial evidence in the record to support the finding of a violation or that the assessment of the penalty by the Administrator constitutes an abuse of discretion.''.

(d) Emergency Powers.--Section 1431(b) (42 U.S.C. 300i(b)) is amended by striking ``$5,000'' and inserting ``$15,000''.

Sec. 114. Public Notification

(a) Public Water Systems.--Section 1414(c) (42 U.S.C. 300g-3(c)) is amended to read as follows:

``(c) Notice to Persons Served.--

``(1) In general.--Each owner or operator of a public water system shall give notice of each of the following to the persons served by the system:

``(A) Notice of any failure on the part of the public water system to--

``(i) comply with an applicable maximum contaminant level or treatment technique requirement of, or a testing procedure prescribed by, a national primary drinking water regulation; or

``(ii) perform monitoring required by section 1445(a).

``(B) If the public water system is subject to a variance granted under subsection (a)(1)(A), (a)(2), or (e) of section 1415 for an inability to meet a maximum contaminant level requirement or is subject to an exemption granted under section 1416, notice of--

``(i) the existence of the variance or exemption; and

``(ii) any failure to comply with the requirements of any schedule prescribed pursuant to the variance or exemption.

``(C) Notice of the concentration level of any unregulated contaminant for which the Administrator has required public notice pursuant to paragraph (2)(E).

``(2) Form, manner, and frequency of notice.--

``(A) <<NOTE: Regulations.>> In general.--The Administrator shall, by regulation, and after consultation with the States, prescribe the manner, frequency, form, and content for giving notice under this subsection. The regulations shall--

``(i) provide for different frequencies of notice based on the differences between violations that are intermittent or infrequent and violations that are continuous or frequent; and

``(ii) take into account the seriousness of any potential adverse health effects that may be involved.

``(B) State requirements.--

``(i) In general.--A State may, by rule, establish alternative notification requirements--

``(I) with respect to the form and content of notice given under and in a manner in accordance with subparagraph (C); and

``(II) with respect to the form and content of notice given under subparagraph (D).

``(ii) Contents.--The alternative requirements shall provide the same type and amount of information as required pursuant to this subsection and regulations issued under subparagraph (A).

``(iii) Relationship to section 1413.--Nothing in this subparagraph shall be construed or applied to modify the requirements of section 1413.

``(C) Violations with potential to have serious adverse effects on human health.--Regulations issued under subparagraph (A) shall specify notification procedures for each violation by a public water system that has the potential to have serious adverse effects on human health as a result of short-term exposure. Each notice of violation provided under this subparagraph shall--

``(i) be distributed as soon as practicable after the occurrence of the violation, but not later than 24 hours after the occurrence of the violation;

``(ii) provide a clear and readily understandable explanation of--

``(I) the violation;

``(II) the potential adverse effects on human health;

``(III) the steps that the public water system is taking to correct the violation; and

``(IV) the necessity of seeking alternative water supplies until the violation is corrected;

``(iii) be provided to the Administrator or the head of the State agency that has primary enforcement responsibility under section 1413 as soon as practicable, but not later than 24 hours after the occurrence of the violation; and

``(iv) as required by the State agency in general regulations of the State agency, or on a case-by-case basis after the consultation referred to in clause (iii), considering the health risks involved--

``(I) be provided to appropriate broadcast media;

``(II) be prominently published in a newspaper of general circulation serving the area not later than 1 day after distribution of a notice pursuant to clause (i) or the date of publication of the next issue of the newspaper; or

``(III) be provided by posting or door-to-door notification in lieu of notification by means of broadcast media or newspaper.

``(D) Written notice.--

``(i) In general.--Regulations issued under subparagraph (A) shall specify notification procedures for violations other than the violations covered by subparagraph (C). <<NOTE: Reports.>> The procedures shall specify that a public water system shall provide written notice to each person served by the system by notice (I) in the first bill (if any) prepared after the date of occurrence of the violation, (II) in an annual report issued not later than 1 year after the date of occurrence of the violation, or (III) by mail or direct delivery as soon as practicable, but not later than 1 year after the date of occurrence of the violation.

``(ii) Form and manner of notice.--The Administrator shall prescribe the form and manner of the notice to provide a clear and readily understandable explanation of the violation, any potential adverse health effects, and the steps that the system is taking to seek alternative water supplies, if any, until the violation is corrected.

``(E) Unregulated contaminants.--The Administrator may require the owner or operator of a public water system to give notice to the persons served by the system of the concentration levels of an unregulated contaminant required to be monitored under section 1445(a).

``(3) Reports.--

``(A) Annual report by state.--

``(i) In general.--Not later than January 1, 1998, and annually thereafter, each State that has primary enforcement responsibility under section 1413 shall prepare, make readily available to the public, and submit to the Administrator an annual report on violations of national primary drinking water regulations by public water systems in the State, including violations with respect to (I) maximum contaminant levels, (II) treatment requirements, (III) variances and exemptions, and (IV) monitoring requirements determined to be significant by the Administrator after consultation with the States.

``(ii) <<NOTE: Publication.>> Distribution.-- The State shall publish and distribute summaries of the report and indicate where the full report is available for review.

``(B) <<NOTE: Native Americans.>> Annual report by administrator.--Not later than July 1, 1998, and annually thereafter, the Administrator shall prepare and make available to the public an annual report summarizing and evaluating reports submitted by States pursuant to subparagraph (A) and notices submitted by public water systems serving Indian Tribes provided to the Administrator pursuant to subparagraph (C) or (D) of paragraph (2) and making recommendations concerning the resources needed to improve compliance with this title. The report shall include information about public water system compliance on Indian reservations and about enforcement activities undertaken and financial assistance provided by the Administrator on Indian reservations, and shall make specific recommendations concerning the resources needed to improve compliance with this title on Indian reservations.

``(4) Consumer confidence reports by community water systems.--

``(A) <<NOTE: Regulations.>> Annual reports to consumers.--The Administrator, in consultation with public water systems, environmental groups, public interest groups, risk communication experts, and the States, and other interested parties, shall issue regulations within 24 months after the date of enactment of this paragraph to require each community water system to mail to each customer of the system at least once annually a report on the level of contaminants in the drinking water purveyed by that system (referred to in this paragraph as a `consumer confidence report'). Such regulations shall provide a brief and plainly worded definition of the terms `maximum contaminant level goal', `maximum contaminant level', `variances', and `exemptions' and brief statements in plain language regarding the health concerns that resulted in regulation of each regulated contaminant. The regulations shall also include a brief and plainly worded explanation regarding contaminants that may reasonably be expected to be present in drinking water, including bottled water. The regulations shall also provide for an Environmental Protection Agency toll-free hotline that consumers can call for more information and explanation.

``(B) Contents of report.--The consumer confidence reports under this paragraph shall include, but not be limited to, each of the following:

``(i) Information on the source of the water purveyed.

``(ii) A brief and plainly worded definition of the terms `maximum contaminant level goal', `maximum contaminant level', `variances', and `exemptions' as provided in the regulations of the Administrator.

``(iii) If any regulated contaminant is detected in the water purveyed by the public water system, a statement setting forth (I) the maximum contaminant level goal, (II) the maximum contaminant level, (III) the level of such contaminant in such water system, and (IV) for any regulated contaminant for which there has been a violation of the maximum contaminant level during the year concerned, the brief statement in plain language regarding the health concerns that resulted in regulation of such contaminant, as provided by the Administrator in regulations under subparagraph (A).

``(iv) Information on compliance with national primary drinking water regulations, as required by the Administrator, and notice if the system is operating under a variance or exemption and the basis on which the variance or exemption was granted.

``(v) Information on the levels of unregulated contaminants for which monitoring is required under section 1445(a)(2) (including levels of cryptosporidium and radon where States determine they may be found).

``(vi) A statement that the presence of contaminants in drinking water does not necessarily indicate that the drinking water poses a health risk and that more information about contaminants and potential health effects can be obtained by calling the Environmental Protection Agency hotline.

A public water system may include such additional information as it deems appropriate for public education. The Administrator may, for not more than 3 regulated contaminants other than those referred to in subclause (IV) of clause (iii), require a consumer confidence report under this paragraph to include the brief statement in plain language regarding the health concerns that resulted in regulation of the contaminant or contaminants concerned, as provided by the Administrator in regulations under subparagraph (A).

``(C) Coverage.--The Governor of a State may determine not to apply the mailing requirement of subparagraph (A) to a community water system serving fewer than 10,000 persons. Any such system shall--

``(i) <<NOTE: Newspapers.>> inform, in the newspaper notice required by clause (iii) or by other means, its customers that the system will not be mailing the report as required by subparagraph (A);

``(ii) make the consumer confidence report available upon request to the public; and

``(iii) <<NOTE: Publication.>> publish the report referred to in subparagraph (A) annually in one or more local newspapers serving the area in which customers of the system are located.

``(D) Alternative to publication.--For any community water system which, pursuant to subparagraph (C), is not required to meet the mailing requirement of subparagraph (A) and which serves 500 persons or fewer, the community water system may elect not to comply with clause (i) or (iii) of subparagraph (C). If the community water system so elects, the system shall, at a minimum--

``(i) <<NOTE: Reports.>> prepare an annual consumer confidence report pursuant to subparagraph (B); and

``(ii) provide notice at least once per year to each of its customers by mail, by door-to-door delivery, by posting or by other means authorized by the regulations of the Administrator that the consumer confidence report is available upon request.

``(E) Alternative form and content.--A State exercising primary enforcement responsibility may establish, by rule, after notice and public comment, alternative requirements with respect to the form and content of consumer confidence reports under this paragraph.''.

(b) <<NOTE: Publication. 21 USC 349 note.>> Bottled Water Study.--

Not later than 18 months after the date of enactment of this Act, the Administrator of the Food and Drug Administration, in consultation with the Administrator of the Environmental Protection Agency, shall publish for public notice and comment a draft study on the feasibility of appropriate methods, if any, of informing customers of the contents of bottled water. The Administrator of the Food and Drug Administration shall publish a final study not later than 30 months after the date of enactment of this Act.

Sec. 115. Variances

The second sentence of section 1415(a)(1)(A) (42 U.S.C. 300g-4(a)(1)(A)) is amended--

(1) by striking ``only be issued to a system after the system's application of'' and inserting ``be issued to a system on condition that the system install''; and

(2) by inserting before the period at the end the following: ``, and based upon an evaluation satisfactory to the State that indicates that alternative sources of water are not reasonably available to the system''.

Sec. 116. Small Systems Variances

Section 1415 (42 U.S.C. 300g-4) is amended by adding at the end the following:

``(e) Small System Variances.--

``(1) In general.--A State exercising primary enforcement responsibility for public water systems under section 1413 (or the Administrator in nonprimacy States) may grant a variance under this subsection for compliance with a requirement specifying a maximum contaminant level or treatment technique contained in a national primary drinking water regulation to--

``(A) public water systems serving 3,300 or fewer persons; and

``(B) with the approval of the Administrator pursuant to paragraph (9), public water systems serving more than 3,300 persons but fewer than 10,000 persons,

if the variance meets each requirement of this subsection.

``(2) Availability of variances.--A public water system may receive a variance pursuant to paragraph (1), if--

``(A) the Administrator has identified a variance technology under section 1412(b)(15) that is applicable to the size and source water quality conditions of the public water system;

``(B) the public water system installs, operates, and maintains, in accordance with guidance or regulations issued by the Administrator, such treatment technology, treatment technique, or other means; and

``(C) the State in which the system is located determines that the conditions of paragraph (3) are met.

``(3) Conditions for granting variances.--A variance under this subsection shall be available only to a system--

``(A) that cannot afford to comply, in accordance with affordability criteria established by the Administrator (or the State in the case of a State that has primary enforcement responsibility under section 1413), with a national primary drinking water regulation, including compliance through--

``(i) treatment;

``(ii) alternative source of water supply; or

``(iii) restructuring or consolidation (unless the Administrator (or the State in the case of a State that has primary enforcement responsibility under section 1413) makes a written determination that restructuring or consolidation is not practicable); and

``(B) for which the Administrator (or the State in the case of a State that has primary enforcement responsibility under section 1413) determines that the terms of the variance ensure adequate protection of human health, considering the quality of the source water for the system and the removal efficiencies and expected useful life of the treatment technology required by the variance.

``(4) Compliance schedules.--A variance granted under this subsection shall require compliance with the conditions of the variance not later than 3 years after the date on which the variance is granted, except that the Administrator (or the State in the case of a State that has primary enforcement responsibility under section 1413) may allow up to 2 additional years to comply with a variance technology, secure an alternative source of water, restructure or consolidate if the Administrator (or the State) determines that additional time is necessary for capital improvements, or to allow for financial assistance provided pursuant to section 1452 or any other Federal or State program.

``(5) <<NOTE: Review.>> Duration of variances.--The Administrator (or the State in the case of a State that has primary enforcement responsibility under section 1413) shall review each variance granted under this subsection not less often than every 5 years after the compliance date established in the variance to determine whether the system remains eligible for the variance and is conforming to each condition of the variance.

``(6) Ineligibility for variances.--A variance shall not be available under this subsection for--

``(A) any maximum contaminant level or treatment technique for a contaminant with respect to which a national primary drinking water regulation was promulgated prior to January 1, 1986; or

``(B) a national primary drinking water regulation for a microbial contaminant (including a bacterium, virus, or other organism) or an indicator or treatment technique for a microbial contaminant.

``(7) Regulations and guidance.--

``(A) In general.--Not later than 2 years after the date of enactment of this subsection and in consultation with the States, the Administrator shall promulgate regulations for variances to be granted under this subsection. The regulations shall, at a minimum, specify--

``(i) procedures to be used by the Administrator or a State to grant or deny variances, including requirements for notifying the Administrator and consumers of the public water system that a variance is proposed to be granted (including information regarding the contaminant and variance) and requirements for a public hearing on the variance before the variance is granted;

``(ii) requirements for the installation and proper operation of variance technology that is identified (pursuant to section 1412(b)(15)) for small systems and the financial and technical capability to operate the treatment system, including operator training and certification;

``(iii) eligibility criteria for a variance for each national primary drinking water regulation, including requirements for the quality of the source water (pursuant to section 1412(b)(15)(A)); and

``(iv) information requirements for variance applications.

``(B) <<NOTE: Publication.>> Affordability criteria.--Not later than 18 months after the date of enactment of the Safe Drinking Water Act Amendments of 1996, the Administrator, in consultation with the States and the Rural Utilities Service of the Department of Agriculture, shall publish information to assist the States in developing affordability criteria. The affordability <<NOTE: Review.>> criteria shall be reviewed by the

States not less often than every 5 years to determine if changes are needed to the criteria.

``(8) Review by the administrator.--

``(A) In general.--The Administrator shall periodically review the program of each State that has primary enforcement responsibility for public water systems under section 1413 with respect to variances to determine whether the variances granted by the State comply with the requirements of this subsection. With respect to affordability, the determination of the Administrator shall be limited to whether the variances granted by the State comply with the affordability criteria developed by the State.

``(B) Notice and publication.--If the Administrator determines that variances granted by a State are not in compliance with affordability criteria developed by the State and the requirements of this subsection, the Administrator shall notify the State in writing of the deficiencies and make public the determination.

``(9) Approval of variances.--A State proposing to grant a variance under this subsection to a public water system serving more than 3,300 and fewer than 10,000 persons shall submit the variance to the Administrator for review and approval prior to the issuance of the variance. The Administrator shall approve the variance if it meets each of the requirements of this subsection. The Administrator shall approve or disapprove the variance within 90 days. If the <<NOTE: Notification.>> Administrator disapproves a variance under this paragraph, the Administrator shall notify the State in writing of the reasons for disapproval and the variance may be resubmitted with modifications to address the objections stated by the Administrator.

``(10) Objections to variances.--

``(A) By the administrator.--The Administrator may review and object to any variance proposed to be granted by a State, if the objection is communicated to the State not later than 90 days after the State proposes to grant the variance. <<NOTE: Notification.>> If the Administrator objects to the granting of a variance, the Administrator shall notify the State in writing of each basis for the objection and propose a modification to the variance to resolve the concerns of the Administrator. The State shall make the recommended modification or respond in writing to each objection. If the State issues the variance without resolving the concerns of the Administrator, the Administrator may overturn the State decision to grant the variance if the Administrator determines that the State decision does not comply with this subsection.

``(B) Petition by consumers.--Not later than 30 days after a State exercising primary enforcement responsibility for public water systems under section 1413 proposes to grant a variance for a public water system, any person served by the system may petition the Administrator to object to the granting of a variance. The Administrator shall respond to the petition and determine whether to object to the variance under subparagraph (A) not later than 60 days after the receipt of the petition.

``(C) Timing.--No variance shall be granted by a State until the later of the following:

``(i) 90 days after the State proposes to grant a variance.

``(ii) If the Administrator objects to the variance, the date on which the State makes the recommended modifications or responds in writing to each objection.''.

Sec. 117. Exemptions

(a) In General.--Section 1416 (42 U.S.C. 300g-5) is amended as follows:

(1) In subsection (a)(1)--

(A) by inserting after ``(which may include economic factors'' the following: ``, including qualification of the public water system as a system serving a disadvantaged community pursuant to section 1452(d)''; and

(B) by inserting after ``treatment technique requirement,'' the following: ``or to implement measures to develop an alternative source of water supply,''.

(2) In subsection (a), by striking ``and'' at the end of paragraph (2), striking the period at the end of paragraph (3) and inserting ``; and'' and by adding the following at the end thereof:

``(4) management or restructuring changes (or both) cannot reasonably be made that will result in compliance with this title or, if compliance cannot be achieved, improve the quality of the drinking water.''.

(3) In subsection (b)(1)(A)--

(A) by striking ``(including increments of progress)'' and inserting ``(including increments of progress or measures to develop an alternative source of water supply)''; and

(B) by striking ``requirement and treatment'' and inserting ``requirement or treatment''.

(4) In subsection (b)(2)--

(A) by striking ``(except as provided in subparagraph (B))'' in subparagraph (A) and all that follows through ``3 years after the date of the issuance of the exemption if'' in subparagraph (B) and inserting the following: ``not later than 3 years after the otherwise applicable compliance date established in section 1412(b)(10).

``(B) No exemption shall be granted unless'';

(B) in subparagraph (B)(i), by striking ``within the period of such exemption'' and inserting ``prior to the date established pursuant to section 1412(b)(10)'';

(C) in subparagraph (B)(ii), by inserting after ``such financial assistance'' the following: ``or assistance pursuant to section 1452, or any other Federal or State program is reasonably likely to be available within the period of the exemption'';

(D) in subparagraph (C)--

(i) by striking ``500 service connections'' and inserting ``a population of 3,300''; and

(ii) by inserting ``, but not to exceed a total of 6 years,'' after ``for one or more additional 2-year periods''; and

(E) by adding at the end the following:

``(D) Limitation.--A public water system may not receive an exemption under this section if the system was granted a variance under section 1415(e).''.

(b) Limited Additional Compliance Period.--(1) The State of New York, on a case-by-case basis and after notice and an opportunity of at least 60 days for public comment, may allow an additional period for compliance with the Surface Water Treatment Rule established pursuant to section 1412(b)(7)(C) of the Safe Drinking Water Act in the case of unfiltered systems in Essex, Columbia, Greene, Dutchess, Rensselaer, Schoharie, Saratoga, Washington, and Warren Counties serving a population of less than 5,000, which meet appropriate disinfection requirements and have adequate watershed protections, so long as the State determines that the public health will be protected during the duration of the additional compliance period and the system agrees to implement appropriate control measures as determined by the State.

(2) <<NOTE: Expiration.>> The additional compliance period referred to in paragraph (1) shall expire on the earlier of the date 3 years after the date on which the Administrator identifies appropriate control technology for the Surface Water Treatment Rule for public water systems in the category that includes such system pursuant to section 1412(b)(4)(E) of

the Safe Drinking Water Act or 5 years after the date of enactment of the Safe Drinking Water Act Amendments of 1996.

Sec. 118. Lead Plumbing and Pipes

Section 1417 (42 U.S.C. 300g-6) is amended as follows:

(1) In subsection (a), by striking paragraph (1) and inserting the following:

``(1) Prohibitions.--

``(A) In general.--No person may use any pipe, any pipe or plumbing fitting or fixture, any solder, or any flux, after June 19, 1986, in the installation or repair of--

``(i) any public water system; or

``(ii) any plumbing in a residential or nonresidential facility providing water for human consumption,

that is not lead free (within the meaning of subsection (d)).

``(B) Leaded joints.--Subparagraph (A) shall not apply to leaded joints necessary for the repair of cast iron pipes.''.

(2) In subsection (a)(2)(A), by inserting ``owner or operator of a'' after ``Each''.

(3) By adding at the end of subsection (a) the following:

``(3) <<NOTE: Effective date.>> Unlawful acts.--Effective 2 years after the date of enactment of this paragraph, it shall be unlawful--

``(A) for any person to introduce into commerce any pipe, or any pipe or plumbing fitting or fixture, that is not lead free, except for a pipe that is used in manufacturing or industrial processing;

``(B) for any person engaged in the business of selling plumbing supplies, except manufacturers, to sell solder or flux that is not lead free; or

``(C) for any person to introduce into commerce any solder or flux that is not lead free unless the solder or flux bears a prominent label stating that it is illegal to use the solder or flux in the installation or repair of any plumbing providing water for human consumption.''.

(4) In subsection (d)--

(A) by striking ``lead, and'' in paragraph (1) and inserting ``lead;'';

(B) by striking ``lead.'' in paragraph (2) and inserting ``lead; and''; and

(C) by adding at the end the following:

``(3) when used with respect to plumbing fittings and fixtures, refers to plumbing fittings and fixtures in compliance with standards established in accordance with subsection (e).''.

(5) By adding at the end the following:

``(e) Plumbing Fittings and Fixtures.--

``(1) In general.--The Administrator shall provide accurate and timely technical information and assistance to qualified third-party certifiers in the development of voluntary standards and testing protocols for the leaching of lead from new plumbing fittings and fixtures that are intended by the manufacturer to dispense water for human ingestion.

``(2) Standards.--

``(A) In general.--If a voluntary standard for the leaching of lead is not established by the date that is 1 year after the date of enactment of this subsection, the Administrator shall, not later than 2 years after the date of enactment of this subsection, promulgate regulations setting a health-effects-based performance standard establishing maximum leaching levels from new plumbing fittings and fixtures that are intended by the manufacturer to dispense water for human ingestion. <<NOTE: Effective date.>> The standard shall become effective on the date that is 5 years after the date of promulgation of the standard.

``(B) Alternative requirement.--If regulations are required to be promulgated under subparagraph (A) and have not been promulgated by the date that is 5 years after the date of enactment of this subsection, no person may import, manufacture, process, or distribute in commerce a new plumbing fitting or fixture, intended by the manufacturer to dispense water for human ingestion, that contains more than 4 percent lead by dry weight.''.

Sec. 119. Capacity Development

Part B (42 U.S.C. 300g et seq.) is amended by adding after section 1419 the following:

``Sec. <<NOTE: 42 USC 300g-9.>> 1420. (a) State Authority for New Systems.--A State shall receive only 80 percent of the allotment that the State is otherwise entitled to receive under section 1452 (relating to State loan funds) unless the State has obtained the legal authority or other means to ensure that all new community water systems and new nontransient, noncommunity water systems commencing operation after October 1, 1999, demonstrate technical, managerial, and financial capacity with respect to each national primary drinking water regulation in effect, or likely to be in effect, on the date of commencement of operations.

``(b) Systems in Significant Noncompliance.--

``(1) List.--Beginning not later than 1 year after the date of enactment of this section, each State shall prepare, periodically update, and submit to the Administrator a list of community water systems and nontransient, noncommunity water systems that have a history of significant noncompliance with this title (as defined in guidelines issued prior to the date of enactment of this section or any revisions of the guidelines that have been made in consultation with the States) and, to the extent practicable, the reasons for noncompliance.

``(2) Report.--Not later than 5 years after the date of enactment of this section and as part of the capacity development strategy of the State, each State shall report to the Administrator on the success of enforcement mechanisms and initial capacity development efforts in assisting the public water systems listed under paragraph (1) to improve technical, managerial, and financial capacity.

``(3) Withholding.--The list and report under this subsection shall be considered part of the capacity development strategy of the State required under subsection (c) of this section for purposes of the withholding requirements of section 1452(a)(1)(G)(i) (relating to State loan funds).

``(c) Capacity Development Strategy.--

``(1) In general.--Beginning 4 years after the date of enactment of this section, a State shall receive only--

``(A) 90 percent in fiscal year 2001;

``(B) 85 percent in fiscal year 2002; and

``(C) 80 percent in each subsequent fiscal year,

of the allotment that the State is otherwise entitled to receive under section 1452 (relating to State loan funds), unless the State is developing and implementing a strategy to assist public water systems in acquiring and maintaining technical, managerial, and financial capacity.

``(2) Content.--In preparing the capacity development strategy, the State shall consider, solicit public comment on, and include as appropriate--

``(A) the methods or criteria that the State will use to identify and prioritize the public water systems most in need of improving technical, managerial, and financial capacity;

``(B) a description of the institutional, regulatory, financial, tax, or legal factors at the Federal, State, or local level that encourage or impair capacity development;

``(C) a description of how the State will use the authorities and resources of this title or other means to--

``(i) assist public water systems in complying with national primary drinking water regulations;

``(ii) encourage the development of partnerships between public water systems to enhance the technical, managerial, and financial capacity of the systems; and

``(iii) assist public water systems in the training and certification of operators;

``(D) a description of how the State will establish a baseline and measure improvements in capacity with respect to national primary drinking water regulations and State drinking water law; and

``(E) an identification of the persons that have an interest in and are involved in the development and implementation of the capacity development strategy (including all appropriate agencies of Federal, State, and local governments, private and nonprofit public water systems, and public water system customers).

``(3) Report.--Not later than 2 years after the date on which a State first adopts a capacity development strategy under this subsection, and every 3 years thereafter, the head of the State agency that has primary responsibility to carry out this title in the State shall submit to the Governor a report that shall also be available to the public on the efficacy of the strategy and progress made toward improving the technical, managerial, and financial capacity of public water systems in the State.

``(4) Review.--The decisions of the State under this section regarding any particular public water system are not subject to review by the Administrator and may not serve as the basis for withholding funds under section 1452.

``(d) Federal Assistance.--

``(1) In general.--The Administrator shall support the States in developing capacity development strategies.

``(2) Informational assistance.--

``(A) In general.--Not later than 180 days after the date of enactment of this section, the Administrator shall--

``(i) <<NOTE: Review. Publication.>> conduct a review of State capacity development efforts in existence on the date of enactment of this section and publish information to assist States and public water systems in capacity development efforts; and ``(ii) initiate a partnership with States, public water systems, and the public to develop information for States on recommended operator certification requirements.

``(B) Publication of information.--The Administrator shall publish the information developed through the partnership under subparagraph (A)(ii) not later than 18 months after the date of enactment of this section.

``(3) Promulgation of drinking water regulations.--In promulgating a national primary drinking water regulation, the Administrator shall include an analysis of the likely effect of compliance with the regulation on the technical, financial, and managerial capacity of public water systems.

``(4) <<NOTE: Publication.>> Guidance for new systems.--Not later than 2 years after the date of enactment of this section, the Administrator shall publish guidance developed in consultation with the States describing legal authorities and other means to ensure that all new community water systems and new nontransient, noncommunity water systems demonstrate technical, managerial, and financial capacity with respect to national primary drinking water regulations.

``(e) Variances and Exemptions.--Based on information obtained under subsection (c)(3), the Administrator shall, as appropriate, modify regulations concerning variances and exemptions for small public water systems to ensure flexibility in the use of the variances and exemptions. Nothing in this subsection shall be interpreted, construed, or applied to affect or alter the requirements of section 1415 or 1416.

``(f) Small Public Water Systems Technology Assistance Centers.--

``(1) Grant program.--The Administrator is authorized to make grants to institutions of higher learning to establish and operate small public water system technology assistance centers in the United States.

``(2) Responsibilities of the centers.--The responsibilities of the small public water system technology assistance centers established under this subsection shall include the conduct of training and technical assistance relating to the information, performance, and technical needs of small public water systems or public water systems that serve Indian Tribes.

``(3) Applications.--Any institution of higher learning interested in receiving a grant under this subsection shall submit to the Administrator an application in such form and containing such information as the Administrator may require by regulation.

``(4) Selection criteria.--The Administrator shall select recipients of grants under this subsection on the basis of the following criteria:

``(A) The small public water system technology assistance center shall be located in a State that is representative of the needs of the region in which the State is located for addressing the drinking water needs of small and rural communities or Indian Tribes.

``(B) The grant recipient shall be located in a region that has experienced problems, or may reasonably be foreseen to experience problems, with small and rural public water systems.

``(C) The grant recipient shall have access to expertise in small public water system technology management.

``(D) The grant recipient shall have the capability to disseminate the results of small public water system technology and training programs.

``(E) The projects that the grant recipient proposes to carry out under the grant are necessary and appropriate.

``(F) The grant recipient has regional support beyond the host institution.

``(5) Consortia of states.--At least 2 of the grants under this subsection shall be made to consortia of States with low population densities.

``(6) Authorization of appropriations.--There are authorized to be appropriated to make grants under this subsection $2,000,000 for each of the fiscal years 1997 through 1999, and $5,000,000 for each of the fiscal years 2000 through 2003.

``(g) Environmental Finance Centers.--

``(1) In general.--The Administrator shall provide initial funding for one or more university-based environmental finance centers for activities that provide technical assistance to State and local officials in developing the capacity of public water systems. Any such funds shall be used only for activities that are directly related to this title.

``(2) <<NOTE: Establishment.>> National capacity development clearinghouse.--The Administrator shall establish a national public water system capacity development clearinghouse to receive and disseminate information with respect to developing, improving, and maintaining financial and managerial capacity at public water systems. The Administrator shall ensure that the clearinghouse does not duplicate other federally supported clearinghouse activities.

``(3) Capacity development techniques.--The Administrator may request an environmental finance center funded under paragraph (1) to develop and test managerial, financial, and institutional techniques for capacity development. The techniques may include capacity assessment methodologies, manual and computer based public water system rate models and capital planning models, public water system consolidation procedures, and regionalization models.

``(4) Authorization of appropriations.--There are authorized to be appropriated to carry out this subsection $1,500,000 for each of the fiscal years 1997 through 2003.

``(5) Limitation.--No portion of any funds made available under this subsection may be used for lobbying expenses.''.

Sec. 120. Authorization of Appropriations for Certain Ground Water-Programs

(a) Critical Aquifer Protection.--Section 1427 (42 U.S.C. 300h-6) is amended as follows:

(1) Subsection (b)(1) is amended by striking ``not later than 24 months after the enactment of the Safe Drinking Water Act Amendments of 1986''.

(2) The table in subsection (m) is amended by adding at the end the following:

``1992-2003......................................15,000,000.''.

(b) Wellhead Protection Areas.--The table in section 1428(k) (42 U.S.C. 300h-7(k)) is amended by adding at the end the following:

``1992-2003......................................30,000,000.''.

(c) Underground Injection Control Grant.--The table in section 1443(b)(5) (42 U.S.C. 300j-2(b)(5)) is amended by adding at the end the following:

``1992-2003......................................15,000,000.''.

Sec. 121. Amendments to Section 1442

Section 1442 (42 U.S.C. 300j-1) is amended--

(1) by redesignating paragraph (3) of subsection (b) as paragraph (3) of subsection (d) and moving such paragraph to appear after paragraph (2) of subsection (d);

(2) by striking subsection (b) (as so amended);

(3) by redesignating subparagraph (B) of subsection (a)(2) as subsection (b) and moving such subsection to appear after subsection (a);

(4) in subsection (a)--

(A) by striking paragraph (2) (as so amended) and inserting the following:

``(2) Information and research facilities.--In carrying out this title, the Administrator is authorized to--

``(A) collect and make available information pertaining to research, investigations, and demonstrations with respect to providing a dependably safe supply of drinking water, together with appropriate recommendations in connection with the information; and

``(B) make available research facilities of the Agency to appropriate public authorities, institutions, and individuals engaged in studies and research relating to this title.'';

(B) by striking paragraph (3); and

(C) by redesignating paragraph (11) as paragraph (3) and moving such paragraph to appear before paragraph (4).

Sec. 122. Technical Assistance

Section 1442(e) (42 U.S.C. 300j-1(e)) is amended to read as follows:

``(e) Technical Assistance.--The Administrator may provide technical assistance to small public water systems to enable such systems to achieve and maintain compliance with applicable national primary drinking water regulations. Such assistance may include circuit-rider and multi-State regional technical assistance programs, training, and preliminary engineering evaluations. The Administrator shall ensure that technical assistance pursuant to this subsection is available in each State. <<NOTE: Nonprofit organizations.>> Each nonprofit organization receiving assistance under this subsection shall consult with the State in which the assistance is to be expended or otherwise made available before using assistance to undertake activities to carry out this subsection. <<NOTE: Appropriation authorization.>> There are authorized to be appropriated to the Administrator to be used for such technical assistance $15,000,000 for each of the fiscal years 1997 through 2003. No portion of any State loan fund established under section 1452 (relating to State loan funds) and no portion of any funds made available under this subsection may be used for lobbying expenses. Of the total amount appropriated under this subsection, 3 percent shall be used for technical assistance to public water systems owned or operated by Indian Tribes.''.

Sec. 123. Operator Certification

Part B (42 U.S.C. 300g et seq.) is amended by adding the following after section 1418:

``Sec. 1419. <<NOTE: Federal Register, publication. 42 USC 300g-8.>> (a) Guidelines.--Not later than 30 months after the date of enactment of the Safe Drinking Water Act Amendments of 1996 and in cooperation with the States, the Administrator shall publish guidelines in the Federal Register, after notice and opportunity for comment from interested persons, including States and public water systems, specifying minimum standards for certification (and recertification) of the operators of community and nontransient noncommunity public water systems. Such guidelines shall take into account existing State programs, the complexity of the system, and other factors aimed at providing an effective program at reasonable cost to States and public water systems, taking into account the size of the system.

``(b) State Programs.--Beginning 2 years after the date on which the Administrator publishes guidelines under subsection (a), the Administrator shall withhold 20 percent of the funds a State is otherwise entitled to receive under section 1452 unless the State has adopted and is implementing a program for the certification of operators of community and nontransient noncommunity public water systems that meets the requirements of the guidelines published pursuant to subsection (a) or that has been submitted in compliance with subsection (c) and that has not been disapproved.

``(c) Existing Programs.--For any State exercising primary enforcement responsibility for public water systems or any other State which has an operator certification program, the guidelines under subsection (a) shall allow the State to enforce such program in lieu of the guidelines under subsection (a) if the State submits the program to the Administrator within 18 months after the publication of the guidelines unless the Administrator determines (within 9 months after the State submits the program to the Administrator) that such program is not substantially equivalent to such guidelines. In making this determination, an existing State program shall be presumed to be substantially equivalent to the guidelines, notwithstanding program differences, based on the size of systems or the quality of source water, providing the State program meets the overall public health objectives of the guidelines. If disapproved, the program may be resubmitted within 6 months after receipt of notice of disapproval.

``(d) Expense Reimbursement.--

``(1) In general.--The Administrator shall provide reimbursement for the costs of training, including an appropriate per diem for unsalaried operators, and certification for persons operating systems serving 3,300 persons or fewer that are required to undergo training pursuant to this section.

``(2) State grants.--The reimbursement shall be provided through grants to States with each State receiving an amount sufficient to cover the reasonable costs for training all such operators in the State, as determined by the Administrator, to the extent required by this section. Grants received by a State pursuant to this paragraph shall first be used to provide reimbursement for training and certification costs of persons operating systems serving 3,300 persons or fewer. If a State has reimbursed all such costs, the State may, after notice to the Administrator, use any remaining funds from the grant for any of the other purposes authorized for grants under section 1452.

``(3) Authorization.--There are authorized to be appropriated to the Administrator to provide grants for reimbursement under this section $30,000,000 for each of fiscal years 1997 through 2003.

``(4) Reservation.--If the appropriation made pursuant to paragraph (3) for any fiscal year is not sufficient to satisfy the requirements of paragraph (1), the Administrator shall, prior to any other allocation or reservation, reserve such sums as necessary from the funds appropriated pursuant to section 1452(m) to provide reimbursement for the training and certification costs mandated by this subsection.''.

Sec. 124. Public Water System Supervision Program

Section 1443(a) (42 U.S.C. 300j-2(a)) is amended as follows:

(1) Paragraph (7) is amended to read as follows:

``(7) Authorization.--For the purpose of making grants under paragraph (1), there are authorized to be appropriated $100,000,000 for each of fiscal years 1997 through 2003.''.

(2) By adding at the end the following:

``(8) Reservation of funds by the administrator.--If the Administrator assumes the primary enforcement responsibility of a State public water system supervision program, the Administrator may reserve from funds made available pursuant to this subsection an amount equal to the amount that would otherwise have been provided to the State pursuant to this subsection. The Administrator shall use the funds reserved pursuant to this paragraph to ensure the full and effective administration of a public water system supervision program in the State.

``(9) State loan funds.--

``(A) Reservation of funds.--For any fiscal year for which the amount made available to the Administrator by appropriations to carry out this subsection is less than the amount that the Administrator determines is necessary to supplement funds made available pursuant to paragraph (8) to ensure the full and effective administration of a public water system supervision program in a State, the Administrator may reserve from the funds made available to the State under section 1452 (relating to State loan funds) an amount that is equal to the amount of the shortfall. This paragraph shall not apply to any State not exercising primary enforcement responsibility for public water systems as of the date of enactment of the Safe Drinking Water Act Amendments of 1996.

``(B) Duty of administrator.--If the Administrator reserves funds from the allocation of a State under subparagraph (A), the Administrator shall carry out in the State each of the activities that would be required of the State if the State had primary enforcement authority under section 1413.''.

Sec. 125. Monitoring and Information Gathering

(a) Review of Existing Requirements.--Paragraph (1) of section 1445(a) (42 U.S.C. 300j-4(a)(1)) is amended to read as follows:

``(1)(A) <<NOTE: Records.>> Every person who is subject to any requirement of this title or who is a grantee, shall establish and maintain such records, make such reports, conduct such monitoring, and provide such information as the Administrator may reasonably require by regulation to assist the Administrator in establishing regulations under this title, in determining whether such person has acted or is acting in compliance with this title, in administering any program of financial assistance under this title, in evaluating the health risks of unregulated contaminants, or in advising the public of such risks. In requiring a public water system to monitor under this subsection, the Administrator may take into consideration the system size and the contaminants likely to be found in the system's drinking water.

``(B) Every person who is subject to a national primary drinking water regulation under section 1412 shall provide such information as the Administrator may reasonably require, after consultation with the State in which such person is located if such State has primary enforcement responsibility for public water systems, on a case-by-case basis, to determine whether such person has acted or is acting in compliance with this title.

``(C) Every person who is subject to a national primary drinking water regulation under section 1412 shall provide such information as the Administrator may reasonably require to assist the Administrator in establishing regulations under section 1412 of this title, after consultation with States and suppliers of water. The Administrator may not require under this subparagraph the installation of treatment equipment or process changes, the testing of treatment technology, or the analysis or processing of monitoring samples, except where the Administrator provides the funding for such activities. Before exercising this authority, the Administrator shall first seek to obtain the information by voluntary submission.

``(D) <<NOTE: Regulations.>> The Administrator shall not later than 2 years after the date of enactment of this subparagraph, after consultation with public health experts, representatives of the general public, and officials of State and local governments, review the monitoring requirements for not fewer than 12 contaminants identified by the Administrator, and promulgate any necessary modifications.''.

(b) Monitoring Relief.--Part B is amended by adding the following new section after section 1417 (42 U.S.C. 300g-6):

``Sec. 1418. <<NOTE: 42 USC 300g-7.>> (a) Interim Monitoring Relief Authority.--

``(1) In general.--A State exercising primary enforcement responsibility for public water systems may modify the monitoring requirements for any regulated or unregulated contaminants for which monitoring is required other than microbial contaminants (or indicators thereof), disinfectants and disinfection byproducts or corrosion byproducts for an interim period to provide that any public water system serving 10,000 persons or fewer shall not be required to conduct additional quarterly monitoring during an interim relief period for such contaminants if--

``(A) monitoring, conducted at the beginning of the period for the contaminant concerned and certified to the State by the public water system, fails to detect the presence of the contaminant in the ground or surface water supplying the public water system; and

``(B) the State, considering the hydrogeology of the area and other relevant factors, determines in writing that the contaminant is unlikely to be detected by further monitoring during such period.

``(2) Termination; timing of monitoring.--The interim relief period referred to in paragraph (1) shall terminate when permanent monitoring relief is adopted and approved for such State, or at the end of 36 months after the date of enactment of the Safe Drinking Water Act Amendments of 1996, whichever comes first. In order to serve as a basis for interim relief, the monitoring conducted at the beginning of the period must occur at the time determined by the State to be the time of the public water system's greatest vulnerability to the contaminant concerned in the relevant ground or surface water, taking into account in the case of pesticides the time of application of the pesticide for the source water area and the travel time for the pesticide to reach such waters and taking into account, in the case of other contaminants, seasonality of precipitation and contaminant travel time.

``(b) Permanent Monitoring Relief Authority.--

``(1) In general.--Each State exercising primary enforcement responsibility for public water systems under this title and having an approved source water assessment program may adopt, in accordance with guidance published by the Administrator, tailored alternative monitoring requirements for public water systems in such State (as an alternative to the monitoring requirements for chemical contaminants set forth in the applicable national primary drinking water regulations) where the State concludes that (based on data available at the time of adoption concerning susceptibility, use, occurrence, or wellhead protection, or from the State's drinking water source water assessment program) such alternative monitoring would provide assurance that it complies with the Administrator's guidelines. The State program must be adequate to assure compliance with, and enforcement of, applicable national primary drinking water regulations. Alternative monitoring shall not apply to regulated microbiological contaminants (or indicators thereof), disinfectants and disinfection byproducts, or corrosion byproducts. The preceding sentence is not intended to limit other authority of the Administrator under other provisions of this title to grant monitoring flexibility.

``(2) Guidelines.--

``(A) In general.--The Administrator shall issue, after notice and comment and at the same time as guidelines are issued for source water assessment under section 1453, guidelines for States to follow in proposing alternative monitoring requirements under paragraph (1) for chemical contaminants. <<NOTE: Federal Register, publication.>> The

Administrator shall publish such guidelines in the Federal Register. The guidelines shall assure that the public health will be protected from drinking water contamination. The guidelines shall require that a State alternative monitoring program apply on a contaminant-by-contaminant basis and that, to be eligible for such alternative monitoring program, a public water system must show the State that the contaminant is not present in the drinking water supply or, if present, it is reliably and consistently below the maximum contaminant level.

``(B) Definition.--For purposes of subparagraph (A), the phrase `reliably and consistently below the maximum contaminant level' means that, although contaminants have been detected in a water supply, the State has sufficient knowledge of the contamination source and extent of contamination to predict that the maximum contaminant level will not be exceeded. In determining that a contaminant is reliably and consistently below the maximum contaminant level, States shall consider the quality and completeness of data, the length of time covered and the volatility or stability of monitoring results during that time, and the proximity of such results to the maximum contaminant level. Wide variations in the analytical results, or analytical results close to the maximum contaminant level, shall not be considered to be reliably and consistently below the maximum contaminant level.

``(3) Effect of detection of contaminants.--The guidelines issued by the Administrator under paragraph (2) shall require that if, after the monitoring program is in effect and operating, a contaminant covered by the alternative monitoring program is detected at levels at or above the maximum contaminant level or is no longer reliably or consistently below the maximum contaminant level, the public water system must either--

``(A) demonstrate that the contamination source has been removed or that other action has been taken to eliminate the contamination problem; or

``(B) test for the detected contaminant pursuant to the applicable national primary drinking water regulation.

``(4) States not exercising primary enforcement responsibility.--The Governor of any State not exercising primary enforcement responsibility under section 1413 on the date of enactment of this section may submit to the Administrator a request that the Administrator modify the monitoring requirements established by the Administrator and applicable to public water systems in that State. After consultation with the Governor, the Administrator shall modify the requirements for public water systems in that State if the request of the Governor is in accordance with each of the requirements of this subsection that apply to alternative monitoring requirements established by States that have primary enforcement responsibility. A decision by the Administrator to approve a request under this clause shall be for a period of 3 years and may subsequently be extended for periods of 5 years.

``(c) Treatment as NPDWR.--All monitoring relief granted by a State to a public water system for a regulated contaminant under subsection (a) or (b) shall be treated as part of the national primary drinking water regulation for that contaminant.

``(d) Other Monitoring Relief.--Nothing in this section shall be construed to affect the authority of the States under applicable national primary drinking water regulations to alter monitoring requirements through waivers or other existing authorities. The Administrator shall periodically review and, as appropriate, revise such authorities.''.

(c) Unregulated Contaminants.--Section 1445(a) (42 U.S.C. 300j-4(a)) is amended by striking paragraphs (2) through (8) and inserting the following:

``(2) Monitoring program for unregulated contaminants.--

``(A) <<NOTE: Regulations.>> Establishment.--The Administrator shall promulgate regulations establishing the criteria for a monitoring program for unregulated contaminants. The regulations shall require monitoring of drinking water supplied by public water systems and shall vary the frequency and schedule for monitoring requirements for systems based on the number of persons served by the system, the source of supply, and the contaminants likely to be found, ensuring that only a representative sample of systems serving 10,000 persons or fewer are required to monitor.

``(B) Monitoring program for certain unregulated contaminants.--

``(i) <<NOTE: Records.>> Initial list.--Not later than 3 years after the date of enactment of the Safe Drinking Water Act Amendments of 1996 and every 5 years thereafter, the Administrator shall issue a list pursuant to subparagraph (A) of not more than 30 unregulated contaminants to be monitored by public water systems and to be included in the national drinking water occurrence data base maintained pursuant to subsection (g).

``(ii) Governors' petition.--The Administrator shall include among the list of contaminants for which monitoring is required under this paragraph each contaminant recommended in a petition signed by the Governor of each of 7 or more States, unless the Administrator determines that the action would prevent the listing of other contaminants of a higher public health concern.

``(C) Monitoring plan for small and medium systems.--

``(i) In general.--Based on the regulations promulgated by the Administrator, each State may develop a representative monitoring plan to assess the occurrence of unregulated contaminants in public water systems that serve a population of 10,000 or fewer in that State. The plan shall require monitoring for systems representative of different sizes, types, and geographic locations in the State.

``(ii) Grants for small system costs.--From funds reserved under section 1452(o) or appropriated under subparagraph (H), the Administrator shall pay the reasonable cost of such testing and laboratory analysis as are necessary to carry out monitoring under the plan.

``(D) Monitoring results.--Each public water system that conducts monitoring of unregulated contaminants pursuant to this paragraph shall provide the results of the monitoring to the primary enforcement authority for the system.

``(E) Notification.--Notification of the availability of the results of monitoring programs required under paragraph (2)(A) shall be given to the persons served by the system.

``(F) Waiver of monitoring requirement.--The Administrator shall waive the requirement for monitoring for a contaminant under this paragraph in a State, if the State demonstrates that the criteria for listing the contaminant do not apply in that State.

``(G) Analytical methods.--The State may use screening methods approved by the Administrator under subsection (i) in lieu of monitoring for particular contaminants under this paragraph.

``(H) Authorization of appropriations.--There are authorized to be appropriated to carry out this paragraph $10,000,000 for each of the fiscal years 1997 through 2003.''.

(d) Screening Methods.--Section 1445 (42 U.S.C. 300j-4) is amended by adding the following after subsection (h):

``(i) Screening Methods.--The Administrator shall review new analytical methods to screen for regulated contaminants and may approve such methods as are more accurate or cost-effective than established reference methods for use in compliance monitoring.''.

Sec. 126. Occurrence Data Base

Section 1445 (42 U.S.C. 300j-4) is amended by adding the following new subsection after subsection (f):

``(g) Occurrence Data Base.--

``(1) In general.--Not later than 3 years after the date of enactment of the Safe Drinking Water Act Amendments of 1996, the Administrator shall assemble and maintain a national drinking water contaminant occurrence data base, using information on the occurrence of both regulated and unregulated contaminants in public water systems obtained under subsection (a)(1)(A) or subsection (a)(2) and reliable information from other public and private sources.

``(2) Public input.--In establishing the occurrence data base, the Administrator shall solicit recommendations from the Science Advisory Board, the States, and other interested parties concerning the development and maintenance of a national drinking water contaminant occurrence data base, including such issues as the structure and design of the data base, data input parameters and requirements, and the use and interpretation of data.

``(3) Use.--The data shall be used by the Administrator in making determinations under section 1412(b)(1) with respect to the occurrence of a contaminant in drinking water at a level of public health concern.

``(4) Public recommendations.--The Administrator shall periodically solicit recommendations from the appropriate officials of the National Academy of Sciences and the States, and any person may submit recommendations to the Administrator, with respect to contaminants that should be included in the national drinking water contaminant occurrence data base, including recommendations with respect to additional unregulated contaminants that should be listed under subsection (a)(2). Any recommendation submitted under this clause shall be accompanied by reasonable documentation that--

``(A) the contaminant occurs or is likely to occur in drinking water; and

``(B) the contaminant poses a risk to public health.

``(5) Public availability.--The information from the data base shall be available to the public in readily accessible form.

``(6) Regulated contaminants.--With respect to each contaminant for which a national primary drinking water regulation has been established, the data base shall include information on the detection of the contaminant at a quantifiable level in public water systems (including detection of the contaminant at levels not constituting a violation of the maximum contaminant level for the contaminant).

``(7) Unregulated contaminants.--With respect to contaminants for which a national primary drinking water regulation has not been established, the data base shall include--

``(A) monitoring information collected by public water systems that serve a population of more than 10,000, as required by the Administrator under subsection (a);

``(B) monitoring information collected from a representative sampling of public water systems that serve a population of 10,000 or fewer; and

``(C) other reliable and appropriate monitoring information on the occurrence of the contaminants in public water systems that is available to the Administrator.''.

Sec. 127. Drinking Water Advisory Council

The second sentence of section 1446(a) <<NOTE: 42 USC 300j-5.>> (42 U.S.C. 300j-6(a)) is amended by inserting before the period at the end the following: ``, of which two such members shall be associated with small, rural public water systems''.

Sec. 128. New York City Watershed Protection Program

Section 1443 (42 U.S.C. 300j-2) is amended by adding at the end the following:

``(d) New York City Watershed Protection Program.--

``(1) In general.--The Administrator is authorized to provide financial assistance to the State of New York for demonstration projects implemented as part of the watershed program for the protection and enhancement of the quality of source waters of the New York City water supply system, including projects that demonstrate, assess, or provide for comprehensive monitoring and surveillance and projects necessary to comply with the criteria for avoiding filtration contained in 40 CFR 141.71. Demonstration projects which shall be eligible for financial assistance shall be certified to the Administrator by the State of New York as satisfying the purposes of this subsection. In certifying projects to the Administrator, the

State of New York shall give priority to monitoring projects that have undergone peer review.

``(2) Report.--Not later than 5 years after the date on which the Administrator first provides assistance pursuant to this paragraph, the Governor of the State of New York shall submit a report to the Administrator on the results of projects assisted.

``(3) Matching requirements.--Federal assistance provided under this subsection shall not exceed 50 percent of the total cost of the protection program being carried out for any particular watershed or ground water recharge area.

``(4) Authorization.--There are authorized to be appropriated to the Administrator to carry out this subsection for

each of fiscal years 1997 through 2003, $15,000,000 for the purpose of providing assistance to the State of New York to carry out paragraph (1).''.

Sec. 129. Federal Agencies

(a) In General.--Section 1447 (42 U.S.C. 300j-6) is amended by redesignating subsection (c) as subsection (d) and by striking subsections (a) and (b) and inserting the following:

``(a) In General.--Each department, agency, and instrumentality of the executive, legislative, and judicial branches of the Federal Government--

``(1) owning or operating any facility in a wellhead protection area;

``(2) engaged in any activity at such facility resulting, or which may result, in the contamination of water supplies in any such area;

``(3) owning or operating any public water system; or

``(4) engaged in any activity resulting, or which may result in, underground injection which endangers drinking water (within the meaning of section 1421(d)(2)),

shall be subject to, and comply with, all Federal, State, interstate, and local requirements, both substantive and procedural (including any requirement for permits or reporting or any provisions for injunctive relief and such sanctions as may be imposed by a court to enforce such relief), respecting the protection of such wellhead areas, respecting such public water systems, and respecting any underground injection in the same manner and to the same extent as any person is subject to such requirements, including the payment of reasonable service charges. The Federal, State, interstate, and local substantive and procedural requirements referred to in this subsection include, but are not limited to, all administrative orders and all civil and administrative penalties and fines, regardless of whether such penalties or fines are punitive or coercive in nature or are imposed for isolated, intermittent, or continuing violations. The United States hereby expressly waives any immunity otherwise applicable to the United States with respect to any such substantive or procedural requirement (including, but not limited to, any injunctive relief, administrative order or civil or administrative penalty or fine referred to in the preceding sentence, or reasonable service charge). The reasonable service charges referred to in this subsection include, but are not limited to, fees or charges assessed in connection with the processing and issuance of permits, renewal of permits, amendments to permits, review of plans, studies, and other documents, and inspection and monitoring of facilities, as well as any other nondiscriminatory charges that are assessed in connection with a Federal, State, interstate, or local regulatory program respecting the protection of wellhead areas or public water systems or respecting any underground injection. Neither the United States, nor any agent, employee, or officer thereof, shall be immune or exempt from any process or sanction of any State or Federal Court with respect to the enforcement of any such injunctive relief. No agent, employee, or officer of the United States shall be personally liable for any civil penalty under any Federal, State, interstate, or local law concerning the protection of wellhead areas or public water systems or concerning underground injection with respect to any act or omission within the scope of the official duties of the agent, employee, or officer.

An agent, employee, or officer of the United States shall be subject to any criminal sanction (including, but not limited to, any fine or imprisonment) under any Federal or State requirement adopted pursuant to this title, but no department, agency, or instrumentality of the executive, legislative, or judicial branch of the Federal Government shall be subject to any such sanction. The President may exempt any facility of any department, agency, or instrumentality in the executive branch from compliance with such a requirement if he determines it to be in the paramount interest of the United States to do so. No such exemption shall be granted due to lack of appropriation unless the President shall have specifically requested such appropriation as a part of the budgetary process and the Congress shall have failed to make available such requested appropriation. Any exemption shall be for a period not in excess of 1 year, but additional exemptions may be granted for periods not to exceed 1 year upon the President's making a new determination. <<NOTE: President. Reports.>> The President shall report each January to the Congress all exemptions from the requirements of this section granted during the preceding calendar year, together with his reason for granting each such exemption.

``(b) Administrative Penalty Orders.--

``(1) In general.--If the Administrator finds that a Federal agency has violated an applicable requirement under this title, the Administrator may issue a penalty order assessing a penalty against the Federal agency.

``(2) Penalties.--The Administrator may, after notice to the agency, assess a civil penalty against the agency in an amount not to exceed $25,000 per day per violation.

``(3) Procedure.--Before an administrative penalty order issued under this subsection becomes final, the Administrator shall provide the agency an opportunity to confer with the Administrator and shall provide the agency notice and an opportunity for a hearing on the record in accordance with chapters 5 and 7 of title 5, United States Code.

``(4) Public review.--

``(A) In general.--Any interested person may obtain review of an administrative penalty order issued under this subsection. The review may be obtained in the United States District Court for the District of Columbia or in the United States District Court for the district in which the violation is alleged to have occurred by the filing of a complaint with the court within the 30-day period beginning on the date the penalty order becomes final. The person filing the complaint shall simultaneously send a copy of the complaint by certified mail to the Administrator and the Attorney General.

``(B) Record.--The Administrator shall promptly file in the court a certified copy of the record on which the order was issued.

``(C) Standard of review.--The court shall not set aside or remand the order unless the court finds that there is not substantial evidence in the record, taken as a whole, to support the finding of a violation or that the assessment of the penalty by the Administrator constitutes an abuse of discretion.

``(D) Prohibition on additional penalties.--The court may not impose an additional civil penalty for a violation that is subject to the order unless the court finds that the assessment constitutes an abuse of discretion by the Administrator.

``(c) Limitation on State Use of Funds Collected From Federal Government.--Unless a State law in effect on the date of enactment of the Safe Drinking Water Act Amendments of 1996 or a State constitution requires the funds to be used in a different manner, all funds collected by a State from the Federal Government from penalties and fines imposed for violation of any substantive or procedural requirement referred to in subsection (a) shall be used by the State only for projects designed to improve or protect the environment or to defray the costs of environmental protection or enforcement.''.

(b) Citizen Enforcement.--(1) The first sentence of section 1449(a) (42 U.S.C. 300j-8(a)) is amended--

(A) in paragraph (1), by striking ``, or'' and inserting a semicolon;

(B) in paragraph (2), by striking the period at the end and inserting ``; or''; and

(C) by adding at the end the following:

``(3) for the collection of a penalty by the United States Government (and associated costs and interest) against any Federal agency that fails, by the date that is 18 months after the effective date of a final order to pay a penalty assessed by the Administrator under section 1429(b), to pay the penalty.''.

(2) Subsection (b) of section 1449 (42 U.S.C. 300j-8(b)) is amended by striking the period at the end of paragraph (2) and inserting ``; or'' and by adding the following new paragraph after paragraph (2):

``(3) under subsection (a)(3) prior to 60 days after the plaintiff has given notice of such action to the Attorney General and to the Federal agency.''.

(c) Washington Aqueduct.--Section 1447 (42 U.S.C. 300j-6) is amended by adding at the end the following:

``(e) Washington Aqueduct.--The Secretary of the Army shall not pass the cost of any penalty assessed under this title on to any customer, user, or other purchaser of drinking water from the Washington Aqueduct system, including finished water from the Dalecarlia or McMillan treatment plant.''.

Sec. 130. State Revolving Loan Funds

Part E (42 U.S.C. 300j et seq.) is amended by adding the following new section after section 1451:

``Sec. 1452. <<NOTE: 42 USC 300j-12.>> (a) General Authority.--

``(1) Grants to states to establish state loan funds.--

``(A) In general.--The Administrator shall offer to enter into agreements with eligible States to make capitalization grants, including letters of credit, to the States under this subsection to further the health protection objectives of this title, promote the efficient use of fund resources, and for other purposes as are specified in this title.

``(B) Establishment of fund.--To be eligible to receive a capitalization grant under this section, a State shall establish a drinking water treatment revolving loan fund (referred to in this section as a `State loan fund') and comply with the other requirements of this section. Each grant to a State under this section shall be deposited in the State loan fund established by the State, except as otherwise provided in this section and in other provisions of this title. No funds authorized by other provisions of this title to be used for other purposes specified in this title shall be deposited in any State loan fund.

``(C) Extended period.--The grant to a State shall be available to the State for obligation during the fiscal year for which the funds are authorized and during the following fiscal year, except that grants made available from funds provided prior to fiscal year 1997 shall be available for obligation during each of the fiscal years 1997 and 1998.

``(D) Allotment formula.--Except as otherwise provided in this section, funds made available to carry out this section shall be allotted to States that have entered into an agreement pursuant to this section (other than the District of Columbia) in accordance with--

``(i) for each of fiscal years 1995 through 1997, a formula that is the same as the formula used to distribute public water system supervision grant funds under section 1443 in fiscal year 1995, except that the minimum proportionate share established in the formula shall be 1 percent of available funds and the formula shall be adjusted to include a minimum proportionate share for the State of Wyoming and the District of Columbia; and

``(ii) for fiscal year 1998 and each subsequent fiscal year, a formula that allocates to each State the proportional share of the State needs identified in the most recent survey conducted pursuant to subsection (h), except that the minimum proportionate share provided to each State shall be the same as the minimum proportionate share provided under clause (i).

``(E) Reallotment.--The grants not obligated by the last day of the period for which the grants are available shall be reallotted according to the appropriate criteria set forth in subparagraph (D), except that the Administrator may reserve and allocate 10 percent of the remaining amount for financial assistance to Indian Tribes in addition to the amount allotted under subsection (i) and none of the funds reallotted by the

Administrator shall be reallotted to any State that has not obligated all sums allotted to the State pursuant to this section during the period in which the sums were available for obligation.

``(F) Nonprimacy states.--The State allotment for a State not exercising primary enforcement responsibility for public water systems shall not be deposited in any such fund but shall be allotted by the Administrator under this subparagraph. Pursuant to section 1443(a)(9)(A) such sums allotted under this subparagraph shall be reserved as needed by the Administrator to exercise primary enforcement responsibility under this title in such State and the remainder shall be reallotted to States exercising primary enforcement responsibility for public water systems for deposit in such funds. Whenever the Administrator makes a final determination pursuant to section 1413(b) that the requirements of section 1413(a) are no longer being met by a State, additional grants for such State under this title shall be immediately terminated by the Administrator. This subparagraph shall not apply to any State not exercising primary enforcement responsibility for public water systems as of the date of enactment of the Safe Drinking Water Act Amendments of 1996.

``(G) Other programs.--

``(i) New system capacity.--Beginning in fiscal year 1999, the Administrator shall withhold 20 percent of each capitalization grant made pursuant to this section to a State unless the State has met the requirements of section 1420(a) (relating to capacity development) and shall withhold 10 percent for fiscal year 2001, 15 percent for fiscal year 2002, and 20 percent for fiscal year 2003 if the State has not complied with the provisions of section 1420(c) (relating to capacity development strategies). Not more than a total of 20 percent of the capitalization grants made to a State in any fiscal year may be withheld under the preceding provisions of this clause. All funds withheld by the Administrator pursuant to this clause shall be reallotted by the Administrator on the basis of the same ratio as is applicable to funds allotted under subparagraph (D). None of the funds reallotted by the Administrator pursuant to this paragraph shall be allotted to a State unless the State has met the requirements of section 1420 (relating to capacity development).

``(ii) Operator certification.--The Administrator shall withhold 20 percent of each capitalization grant made pursuant to this section unless the State has met the requirements of 1419 (relating to operator certification). All funds withheld by the Administrator pursuant to this clause shall be reallotted by the Administrator on the basis of the same ratio as applicable to funds allotted under subparagraph (D). None of the funds reallotted by the Administrator pursuant to this paragraph shall be allotted to a State unless the State has met the requirements of section 1419 (relating to operator certification).

``(2) Use of funds.--Except as otherwise authorized by this title, amounts deposited in a State loan fund, including loan repayments and interest earned on such amounts, shall be used only for providing loans or loan guarantees, or as a source of reserve and security for leveraged loans, the proceeds of which are deposited in a State loan fund established under paragraph (1), or other financial assistance authorized under this section to community water systems and nonprofit noncommunity water systems, other than systems owned by Federal agencies. Financial assistance under this section may be used by a public water system only for expenditures (not including monitoring, operation, and maintenance expenditures) of a type or category which the Administrator has determined, through guidance, will facilitate compliance with national primary drinking water regulations applicable to the system under section 1412 or otherwise significantly further the health protection objectives of this title. The funds may also be used to provide loans to a system referred to in section 1401(4)(B) for the purpose of providing the treatment described in section 1401(4)(B)(i)(III). The funds shall not be used for the acquisition of real property or interests therein, unless the acquisition is integral to a project authorized by this paragraph and the purchase is from a willing seller. Of the amount credited to any State loan fund established under this section in any fiscal year, 15 percent shall be available solely for providing loan assistance to public water systems which regularly serve fewer than 10,000 persons to the extent such funds can be obligated for eligible projects of public water systems.

``(3) Limitation.--

``(A) In general.--Except as provided in subparagraph (B), no assistance under this section shall be provided to a public water system that--

``(i) does not have the technical, managerial, and financial capability to ensure compliance with the requirements of this title; or

``(ii) is in significant noncompliance with any requirement of a national primary drinking water regulation or variance.

``(B) Restructuring.--A public water system described in subparagraph (A) may receive assistance under this section if--

``(i) the use of the assistance will ensure compliance; and

``(ii) if subparagraph (A)(i) applies to the system, the owner or operator of the system agrees to undertake feasible and appropriate changes in operations (including ownership, management, accounting, rates, maintenance, consolidation, alternative water supply, or other procedures) if the State determines that the measures are necessary to ensure that the system has the technical, managerial, and financial capability to comply with the requirements of this title over the long term.

``(C) Review.--Prior to providing assistance under this section to a public water system that is in significant noncompliance with any requirement of a national primary drinking water regulation or variance, the State shall conduct a review to determine whether subparagraph (A)(i) applies to the system.

``(b) Intended Use Plans.--

``(1) In general.--After providing for public review and comment, each State that has entered into a capitalization agreement pursuant to this section shall annually prepare a plan that identifies the intended uses of the amounts available to the State loan fund of the State.

``(2) Contents.--An intended use plan shall include--

``(A) a list of the projects to be assisted in the first fiscal year that begins after the date of the plan, including a description of the project, the expected terms of financial assistance, and the size of the community served;

``(B) the criteria and methods established for the distribution of funds; and

``(C) a description of the financial status of the State loan fund and the short-term and long-term goals of the State loan fund.

``(3) Use of funds.--

``(A) In general.--An intended use plan shall provide, to the maximum extent practicable, that priority for the use of funds be given to projects that--

``(i) address the most serious risk to human health;

``(ii) are necessary to ensure compliance with the requirements of this title (including requirements for filtration); and

``(iii) assist systems most in need on a per household basis according to State affordability criteria.

``(B) <<NOTE: Publications. Records.>> List of projects.--Each State shall, after notice and opportunity for public comment, publish and periodically update a list of projects in the State that are eligible for assistance under this section, including the priority assigned to each project and, to the extent known, the expected funding schedule for each project.

``(c) Fund Management.--Each State loan fund under this section shall be established, maintained, and credited with repayments and interest. The fund corpus shall be available in perpetuity for providing financial assistance under this section. To the extent amounts in the fund are not required for current obligation or expenditure, such amounts shall be invested in interest bearing obligations.

``(d) Assistance for Disadvantaged Communities.--

``(1) Loan subsidy.--Notwithstanding any other provision of this section, in any case in which the State makes a loan pursuant to subsection (a)(2) to a disadvantaged community or to a community that the State expects to become a disadvantaged community as the result of a proposed project, the State may provide additional subsidization (including forgiveness of principal).

``(2) Total amount of subsidies.--For each fiscal year, the total amount of loan subsidies made by a State pursuant to paragraph (1) may not exceed 30 percent of the amount of the capitalization grant received by the State for the year.

``(3) Definition of disadvantaged community.--In this subsection, the term `disadvantaged community' means the service area of a public water system that meets affordability criteria established after public review and comment by the State in which the public water system is located. The Administrator may publish information to assist States in establishing affordability criteria.

``(e) State Contribution.--Each agreement under subsection (a) shall require that the State deposit in the State loan fund from State moneys an amount equal to at least 20 percent of the total amount of the grant to be made to the State on or before the date on which the grant payment is made to the State, except that a State shall not be required to deposit such amount into the fund prior to the date on which each grant payment is made for fiscal years 1994, 1995, 1996, and 1997 if the State deposits the State contribution amount into the State loan fund prior to September 30, 1999.

``(f) Types of Assistance.--Except as otherwise limited by State law, the amounts deposited into a State loan fund under this section may be used only--

``(1) to make loans, on the condition that--

``(A) the interest rate for each loan is less than or equal to the market interest rate, including an interest free loan;

``(B) principal and interest payments on each loan will commence not later than 1 year after completion of the project for which the loan was made, and each loan will be fully amortized not later than 20 years after the completion of the project, except that in the case of a disadvantaged community (as defined in subsection (d)(3)), a State may provide an extended term for a loan, if the extended term--

``(i) terminates not later than the date that is 30 years after the date of project completion; and

``(ii) does not exceed the expected design life of the project;

``(C) the recipient of each loan will establish a dedicated source of revenue (or, in the case of a privately owned system, demonstrate that there is adequate security) for the repayment of the loan; and

``(D) the State loan fund will be credited with all payments of principal and interest on each loan;

``(2) to buy or refinance the debt obligation of a municipality or an intermunicipal or interstate agency within the State at an interest rate that is less than or equal to the market interest rate in any case in which a debt obligation is incurred after July 1, 1993;

``(3) to guarantee, or purchase insurance for, a local obligation (all of the proceeds of which finance a project eligible for assistance under this section) if the guarantee or purchase would improve credit market access or reduce the interest rate applicable to the obligation;

``(4) as a source of revenue or security for the payment of principal and interest on revenue or general obligation bonds issued by the State if the proceeds of the sale of the bonds will be deposited into the State loan fund; and

``(5) to earn interest on the amounts deposited into the State loan fund.

``(g) Administration of State Loan Funds.--

``(1) Combined financial administration.--Notwithstanding subsection (c), a State may (as a convenience and to avoid unnecessary administrative costs) combine, in accordance with State law, the financial administration of a State loan fund established under this section with the financial administration of any other revolving fund established by the State if otherwise not prohibited by the law under which the State loan fund was established and if the Administrator determines that--

``(A) the grants under this section, together with loan repayments and interest, will be separately accounted for and used solely for the purposes specified in subsection (a); and

``(B) the authority to establish assistance priorities and carry out oversight and related activities (other than financial administration) with respect to assistance remains with the State agency having primary responsibility for administration of the State program under section 1413, after consultation with other appropriate State agencies (as determined by the State): Provided, That in nonprimacy States eligible to receive assistance under this section, the Governor shall determine which State agency will have authority to establish priorities for financial assistance from the State loan fund.

``(2) Cost of administering fund.--Each State may annually use up to 4 percent of the funds allotted to the State under this section to cover the reasonable costs of administration of the programs under this section, including the recovery of reasonable costs expended to establish a State loan fund which are incurred after the date of enactment of this section, and to provide technical assistance to public water systems within the State. For fiscal year 1995 and each fiscal year thereafter, each State may use up to an additional 10 percent of the funds allotted to the State under this section--

``(A) for public water system supervision programs under section 1443(a);

``(B) to administer or provide technical assistance through source water protection programs;

``(C) to develop and implement a capacity development strategy under section 1420(c); and

``(D) for an operator certification program for purposes of meeting the requirements of section 1419,

if the State matches the expenditures with at least an equal amount of State funds. At least half of the match must be additional to the amount expended by the State for public water supervision in fiscal year 1993. An additional 2 percent of the funds annually allotted to each State under this section may be used by the State to provide technical assistance to public water systems serving 10,000 or fewer persons in the State. Funds utilized under subparagraph (B) shall not be used for enforcement actions.

``(3) <<NOTE: Publication.>> Guidance and regulations.--The Administrator shall publish guidance and promulgate regulations as may be necessary to carry out the provisions of this section, including--

``(A) provisions to ensure that each State commits and expends funds allotted to the State under this section as efficiently as possible in accordance with this title and applicable State laws;

``(B) guidance to prevent waste, fraud, and abuse; and

``(C) guidance to avoid the use of funds made available under this section to finance the expansion of any public water system in anticipation of future population growth.

The guidance and regulations shall also ensure that the States, and public water systems receiving assistance under this section, use accounting, audit, and fiscal procedures that conform to generally accepted accounting standards.

``(4) State report.--Each State administering a loan fund and assistance program under this subsection shall publish and submit to the Administrator a report every 2 years on its activities under this section, including the findings of the most recent audit of the fund and the entire State allotment. The Administrator shall periodically audit all State loan funds established by, and all other amounts allotted to, the States pursuant to this section in accordance with procedures established by the Comptroller General.

``(h) <<NOTE: Reports.>> Needs Survey.--The Administrator shall conduct an assessment of water system capital improvement needs of all eligible public water systems in the United States and submit a report to the Congress containing the results of the assessment within 180 days after the date of enactment of the Safe Drinking Water Act Amendments of 1996 and every 4 years thereafter.

``(i) Indian Tribes.--

``(1) In general.--1\1/2\ percent of the amounts appropriated annually to carry out this section may be used by the Administrator to make grants to Indian Tribes and Alaska Native villages that have not otherwise received either grants from the Administrator under this section or assistance from State loan funds established under this section. The grants may only be used for expenditures by tribes and villages for public water system expenditures referred to in subsection (a)(2).

``(2) Use of funds.--Funds reserved pursuant to paragraph (1) shall be used to address the most significant threats to public health associated with public water systems that serve Indian Tribes, as determined by the Administrator in consultation with the Director of the Indian Health Service and Indian Tribes.

``(3) Alaska native villages.--In the case of a grant for a project under this subsection in an Alaska Native village, the Administrator is also authorized to make grants to the State of Alaska for the benefit of Native villages. An amount not to exceed 4 percent of the grant amount may be used by the State of Alaska for project management.

``(4) Needs assessment.--The Administrator, in consultation with the Director of the Indian Health Service and Indian Tribes, shall, in accordance with a schedule that is consistent with the needs surveys conducted pursuant to subsection (h), prepare surveys and assess the needs of drinking water treatment facilities to serve Indian Tribes, including an evaluation of the public water systems that pose the most significant threats to public health.

``(j) Other Areas.--Of the funds annually available under this section for grants to States, the Administrator shall make allotments in accordance with section 1443(a)(4) for the Virgin Islands, the Commonwealth of the Northern Mariana Islands, American Samoa, and Guam. The grants allotted as provided in this subsection may be provided by the Administrator to the governments of such areas, to public water systems in such areas, or to both, to be used for the public water system expenditures referred to in subsection (a)(2). The grants, and grants for the District of Columbia, shall not be deposited in State loan funds. The total allotment of grants under this section for all areas described in this subsection in any fiscal year shall not exceed 0.33 percent of the aggregate amount made available to carry out this section in that fiscal year.

``(k) Other Authorized Activities.--

``(1) In general.--Notwithstanding subsection (a)(2), a State may take each of the following actions:

``(A) Provide assistance, only in the form of a loan, to one or more of the following:

``(i) Any public water system described in subsection (a)(2) to acquire land or a conservation easement from a willing seller or grantor, if the purpose of the acquisition is to protect the source water of the system from contamination and to ensure compliance with national primary drinking water regulations.

``(ii) Any community water system to implement local, voluntary source water protection measures to protect source water in areas delineated pursuant to section 1453, in order to facilitate compliance with national primary drinking water regulations applicable to the system under section 1412 or otherwise significantly further the health protection objectives of this title. Funds authorized under this clause may be used to fund only voluntary, incentive-based mechanisms.

``(iii) Any community water system to provide funding in accordance with section 1454(a)(1)(B)(i).

``(B) Provide assistance, including technical and financial assistance, to any public water system as part of a capacity development strategy developed and implemented in accordance with section 1420(c).

``(C) Make expenditures from the capitalization grant of the State for fiscal years 1996 and 1997 to delineate and assess source water protection areas in accordance with section 1453, except that funds set aside for such expenditure shall be obligated within 4 fiscal years.

``(D) Make expenditures from the fund for the establishment and implementation of wellhead protection programs under section 1428.

``(2) Limitation.--For each fiscal year, the total amount of assistance provided and expenditures made by a State under this subsection may not exceed 15 percent of the amount of the capitalization grant received by the State for that year and may not exceed 10 percent of that amount for any one of the following activities:

``(A) To acquire land or conservation easements pursuant to paragraph (1)(A)(i).

``(B) To provide funding to implement voluntary, incentive-based source water quality protection measures pursuant to clauses (ii) and (iii) of paragraph (1)(A).

``(C) To provide assistance through a capacity development strategy pursuant to paragraph (1)(B).

``(D) To make expenditures to delineate or assess source water protection areas pursuant to paragraph (1)(C).

``(E) To make expenditures to establish and implement wellhead protection programs pursuant to paragraph (1)(D).

``(3) Statutory construction.--Nothing in this section creates or conveys any new authority to a State, political subdivision of a State, or community water system for any new regulatory measure, or limits any authority of a State, political subdivision of a State or community water system.

``(l) Savings.--The failure or inability of any public water system to receive funds under this section or any other loan or grant program, or any delay in obtaining the funds, shall not alter the obligation of the system to comply in a timely manner with all applicable drinking water standards and requirements of this title.

``(m) Authorization of Appropriations.--There are authorized to be appropriated to carry out the purposes of this section $599,000,000 for the fiscal year 1994 and $1,000,000,000 for each of the fiscal years 1995 through 2003. To the extent amounts authorized to be appropriated under this subsection in any fiscal year are not appropriated in that fiscal year, such amounts are authorized to be appropriated in a subsequent fiscal year (prior to the fiscal year 2004). Such sums shall remain available until expended.

``(n) Health Effects Studies.--From funds appropriated pursuant to this section for each fiscal year, the Administrator shall reserve $10,000,000 for health effects studies on drinking water contaminants authorized by the Safe Drinking Water Act Amendments of 1996. In allocating funds made available under this subsection, the Administrator shall give priority to studies concerning the health effects of cryptosporidium (as authorized by section 1458(c)), disinfection byproducts (as authorized by section 1458(c)), and arsenic (as authorized by section 1412(b)(12)(A)), and the implementation of a plan for studies of subpopulations at greater risk of adverse effects (as authorized by section 1458(a)).

``(o) Monitoring for Unregulated Contaminants.--From funds appropriated pursuant to this section for each fiscal year beginning with fiscal year 1998, the Administrator shall reserve $2,000,000 to pay the costs of monitoring for unregulated contaminants under section 1445(a)(2)(C).

``(p) Demonstration Project for State of Virginia.--Notwithstanding the other provisions of this section limiting the use of funds deposited in a State loan fund from any State allotment, the State of Virginia may, as a single demonstration and with the approval of the Virginia General Assembly and the Administrator, conduct a program to demonstrate alternative approaches to intergovernmental coordination to assist in the financing of new drinking water facilities in the following rural communities in southwestern Virginia where none exists on the date of enactment of the Safe Drinking Water Act Amendments of 1996 and where such communities are experiencing economic hardship: Lee County, Wise County, Scott County, Dickenson County, Russell County, Buchanan County, Tazewell County, and the city of Norton, Virginia. The funds allotted to that State and deposited in the State loan fund may be loaned to a regional endowment fund for the purpose set forth in this subsection under a plan to be approved by the Administrator. The plan may include an advisory group that includes representatives of such counties.

``(q) Small System Technical Assistance.--The Administrator may reserve up to 2 percent of the total funds appropriated pursuant to subsection (m) for each of the fiscal years 1997 through 2003 to carry out the provisions of section 1442(e) (relating to technical assistance for small systems), except that the total amount of funds made available for such purpose in any fiscal year through appropriations (as authorized by section 1442(e)) and reservations made pursuant to this subsection shall not exceed the amount authorized by section 1442(e).

``(r) Evaluation.--The Administrator shall conduct an evaluation of the effectiveness of the State loan funds through fiscal year 2001. The evaluation shall be submitted to the Congress at the same time as the President submits to the Congress, pursuant to section 1108 of title 31, United States Code, an appropriations request for fiscal year 2003 relating to the budget of the Environmental Protection Agency.''.

Sec. 131. State Ground Water Protection Grants

Part C (42 U.S.C. 300h et seq.) is amended by adding at the end the following:

``Sec. 1429. <<NOTE: Publication. Regulations. 42 USC 300h-8.>> (a) In General.--The Administrator may make a grant to a State for the development and implementation of a State program to ensure the coordinated and comprehensive protection of ground water resources within the State.

``(b) Guidance.--Not later than 1 year after the date of enactment of the Safe Drinking Water Act Amendments of 1996, and annually thereafter, the Administrator shall publish guidance that establishes procedures for application for State ground water protection program assistance and that identifies key elements of State ground water protection programs.

``(c) Conditions of Grants.--

``(1) In general.--The Administrator shall award grants to States that submit an application that is approved by the Administrator. The Administrator shall determine the amount of a grant awarded pursuant to this paragraph on the basis of an assessment of the extent of ground water resources in the State and the likelihood that awarding the grant will result in sustained and reliable protection of ground water quality.

``(2) Innovative program grants.--The Administrator may also award a grant pursuant to this subsection for innovative programs proposed by a State for the prevention of ground water contamination.

``(3) Allocation of funds.--The Administrator shall, at a minimum, ensure that, for each fiscal year, not less than 1 percent of funds made available to the Administrator by appropriations to carry out this section are allocated to each State that submits an application that is approved by the Administrator pursuant to this section.

``(4) Limitation on grants.--No grant awarded by the Administrator may be used for a project to remediate ground water contamination.

``(d) Amount of Grants.--The amount of a grant awarded pursuant to paragraph (1) shall not exceed 50 percent of the eligible costs of carrying out the ground water protection program that is the subject of the grant (as determined by the Administrator) for the 1-year period beginning on the date that the grant is awarded. The State shall pay a State share to cover the costs of the ground water protection program from State funds in an amount that is not less than 50 percent of the cost of conducting the program.

``(e) Evaluations and Reports.--Not later than 3 years after the date of enactment of the Safe Drinking Water Act Amendments of 1996, and every 3 years thereafter, the Administrator shall evaluate the State ground water protection programs that are the subject of grants awarded pursuant to this section and report to the Congress on the status of ground water quality in the United States and the effectiveness of State programs for ground water protection.

``(f) Authorization of Appropriations.--There are authorized to be appropriated to carry out this section $15,000,000 for each of fiscal years 1997 through 2003.".

Sec. 132. Source Water Assessment

(a) In General.--Part E (42 U.S.C. 300j et seq.) is amended by adding at the end the following:

``Sec. 1453. <<NOTE: 42 USC 300j-13.>> (a) Source Water Assessment.--

``(1) <<NOTE: Publication.>> Guidance.--Within 12 months after the date of enactment of the Safe Drinking Water Act Amendments of 1996, after notice and comment, the Administrator shall publish guidance for States exercising primary enforcement responsibility for public water systems to carry out directly or through delegation (for the protection and benefit of public water systems and for the support of monitoring flexibility) a source water assessment program within the State's boundaries. Each State adopting modifications to monitoring requirements pursuant to section 1418(b) shall, prior to adopting such modifications, have an approved source water assessment program under this section and shall carry out the program either directly or through delegation.

``(2) Program requirements.--A source water assessment program under this subsection shall--

``(A) delineate the boundaries of the assessment areas in such State from which one or more public water systems in the State receive supplies of drinking water, using all reasonably available hydrogeologic information on the sources of the supply of drinking water in the State and the water flow, recharge, and discharge and any other reliable information as the State deems necessary to adequately determine such areas; and

``(B) identify for contaminants regulated under this title for which monitoring is required under this title (or any unregulated contaminants selected by the State, in its discretion, which the State, for the purposes of this subsection, has determined may present a threat to public health), to the extent practical, the origins within each delineated area of such contaminants to determine the susceptibility of the public water systems in the delineated area to such contaminants.

``(3) Approval, implementation, and monitoring relief.--A State source water assessment program under this subsection shall be submitted to the Administrator within 18 months after the Administrator's guidance is issued under this subsection and shall be deemed approved 9 months after the date of such submittal unless the Administrator disapproves the program as provided in section 1428(c). States shall begin implementation of the program immediately after its approval. The Administrator's approval of a State program under this subsection shall include a timetable, established in consultation with the State, allowing not more than 2 years for completion after approval of the program. Public water systems seeking monitoring relief in addition to the interim relief provided under section 1418(a) shall be eligible for monitoring relief, consistent with section 1418(b), upon completion of the assessment in the delineated source water assessment area or areas concerned.

``(4) Timetable.--The timetable referred to in paragraph (3) shall take into consideration the availability to the State of funds under section 1452 (relating to State loan funds) for assessments and other relevant factors. The Administrator may extend any timetable included in a State program approved under paragraph (3) to extend the period for completion by an additional 18 months.

``(5) Demonstration project.--The Administrator shall, as soon as practicable, conduct a demonstration project, in consultation with other Federal agencies, to demonstrate the most effective and protective means of assessing and protecting source waters serving large metropolitan areas and located on Federal lands.

``(6) Use of other programs.--To avoid duplication and to encourage efficiency, the program under this section may make use of any of the following:

``(A) Vulnerability assessments, sanitary surveys, and monitoring programs.

``(B) Delineations or assessments of ground water sources under a State wellhead protection program developed pursuant to this section.

``(C) Delineations or assessments of surface or ground water sources under a State pesticide management plan developed pursuant to the Pesticide and Ground Water State Management Plan Regulation (subparts I and J of part 152 of title 40, Code of Federal Regulations), promulgated under section 3(d) of the Federal Insecticide, Fungicide, and Rodenticide Act (7 U.S.C. 136a(d)).

``(D) Delineations or assessments of surface water sources under a State watershed initiative or to satisfy the watershed criterion for determining if filtration is required under the Surface Water Treatment Rule (section 141.70 of title 40, Code of Federal Regulations).

``(E) Delineations or assessments of surface or ground water sources under programs or plans pursuant to the Federal Water Pollution Control Act.

``(7) Public availability.--The State shall make the results of the source water assessments conducted under this subsection available to the public.

``(b) Approval and Disapproval.--For provisions relating to program approval and disapproval, see section 1428(c).''.

(b) Approval and Disapproval of State Programs.--Section 1428 (42 U.S.C. 300h-7) is amended as follows:

(1) Amend the first sentence of subsection (c)(1) to read as follows: ``If, in the judgment of the Administrator, a State program or portion thereof under subsection (a) is not adequate to protect public water systems as required by subsection (a) or a State program under section 1453 or section 1418(b) does not meet the applicable requirements of section 1453 or section 1418(b), the Administrator shall disapprove such program or portion thereof.''.

(2) Add after the second sentence of subsection (c)(1) the following: ``A State program developed pursuant to section 1453 or section 1418(b) shall be deemed to meet the applicable requirements of section 1453 or section 1418(b) unless the

Administrator determines within 9 months of the receipt of the program that such program (or portion thereof) does not meet such requirements.''.

(3) In the third sentence of subsection (c)(1) and in subsection (c)(2), strike ``is inadequate'' and insert ``is disapproved''.

(4) In subsection (b), add the following before the period at the end of the first sentence: ``and source water assessment programs under section 1453''.

Sec. 133. Source Water Petition Program

(a) In General.--Part E (42 U.S.C. 300j et seq.) is amended by adding at the end the following:

``Sec. 1454. <<NOTE: 42 USC 300j-14.>> (a) Petition Program.--

``(1) In general.--

``(A) Establishment.--A State may establish a program under which an owner or operator of a community water system in the State, or a municipal or local government or political subdivision of a State, may submit a source water quality protection partnership petition to the State requesting that the State assist in the local development of a voluntary, incentive-based partnership, among the owner, operator, or government and other persons likely to be affected by the recommendations of the partnership, to--

``(i) reduce the presence in drinking water of contaminants that may be addressed by a petition by considering the origins of the contaminants, including to the maximum extent practicable the specific activities that affect the drinking water supply of a community;

``(ii) obtain financial or technical assistance necessary to facilitate establishment of a partnership, or to develop and implement recommendations of a partnership for the protection of source water to assist in the provision of drinking water that complies with national primary drinking water regulations with respect to contaminants addressed by a petition; and

``(iii) develop recommendations regarding voluntary and incentive-based strategies for the long-term protection of the source water of community water systems.

``(B) Funding.--Each State may--

``(i) use funds set aside pursuant to section 1452(k)(1)(A)(iii) by the State to carry out a program described in subparagraph (A), including assistance to voluntary local partnerships for the development and implementation of partnership recommendations for the protection of source water such as source water quality assessment, contingency plans, and demonstration projects for partners within a source water area delineated under section 1453(a); and

``(ii) provide assistance in response to a petition submitted under this subsection using funds referred to in subsection (b)(2)(B).

``(2) Objectives.--The objectives of a petition submitted under this subsection shall be to--

``(A) facilitate the local development of voluntary, incentive-based partnerships among owners and operators of community water systems, governments, and other persons in source water areas; and

``(B) obtain assistance from the State in identifying resources which are available to implement the recommendations of the partnerships to address the origins of drinking water contaminants that may be addressed by a petition (including to the maximum extent practicable the specific activities contributing to the presence of the contaminants) that affect the drinking water supply of a community.

``(3) Contaminants addressed by a petition.--A petition submitted to a State under this subsection may address only those contaminants--

``(A) that are pathogenic organisms for which a national primary drinking water regulation has been established or is required under section 1412; or

``(B) for which a national primary drinking water regulation has been promulgated or proposed and that are detected by adequate monitoring methods in the source water at the intake structure or in any collection, treatment, storage, or distribution facilities by the community water systems at levels--

``(i) above the maximum contaminant level; or

``(ii) that are not reliably and consistently below the maximum contaminant level.

``(4) Contents.--A petition submitted under this subsection shall, at a minimum--

``(A) include a delineation of the source water area in the State that is the subject of the petition;

``(B) identify, to the maximum extent practicable, the origins of the drinking water contaminants that may be addressed by a petition (including to the maximum extent practicable the specific activities contributing to the presence of the contaminants) in the source water area delineated under section 1453;

``(C) identify any deficiencies in information that will impair the development of recommendations by the voluntary local partnership to address drinking water contaminants that may be addressed by a petition;

``(D) specify the efforts made to establish the voluntary local partnership and obtain the participation of--

``(i) the municipal or local government or other political subdivision of the State with jurisdiction over the source water area delineated under section 1453; and

``(ii) each person in the source water area delineated under section 1453--

``(I) who is likely to be affected by recommendations of the voluntary local partnership; and

``(II) whose participation is essential to the success of the partnership;

``(E) outline how the voluntary local partnership has or will, during development and implementation of recommendations of the voluntary local partnership, identify, recognize and take into account any voluntary or other activities already being undertaken by persons in the source water area delineated under section 1453 under Federal or State law to reduce the likelihood that contaminants will occur in drinking water at levels of public health concern; and

``(F) specify the technical, financial, or other assistance that the voluntary local partnership requests of the State to develop the partnership or to implement recommendations of the partnership.

``(b) Approval or Disapproval of Petitions.--

``(1) In general.--After providing notice and an opportunity for public comment on a petition submitted under subsection (a), the State shall approve or disapprove the petition, in whole or in part, not later than 120 days after the date of submission of the petition.

``(2) Approval.--The State may approve a petition if the petition meets the requirements established under subsection (a). The notice of approval shall, at a minimum, include for informational purposes--

``(A) an identification of technical, financial, or other assistance that the State will provide to assist in addressing the drinking water contaminants that may be addressed by a petition based on--

``(i) the relative priority of the public health concern identified in the petition with respect to the other water quality needs identified by the State;

``(ii) any necessary coordination that the State will perform of the program established under this section with programs implemented or planned by other States under this section; and

``(iii) funds available (including funds available from a State revolving loan fund established under title VI of the Federal Water Pollution Control Act (33 U.S.C. 1381 et seq.)) or section 1452;

``(B) a description of technical or financial assistance pursuant to Federal and State programs that is available to assist in implementing recommendations of the partnership in the petition, including-

``(i) any program established under the Federal Water Pollution Control Act (33 U.S.C. 1251 et seq.);

``(ii) the program established under section 6217 of the Coastal Zone Act Reauthorization Amendments of 1990 (16 U.S.C. 1455b);

``(iii) the agricultural water quality protection program established under chapter 2 of subtitle D of title XII of the Food Security Act of 1985 (16 U.S.C. 3838 et seq.);

``(iv) the sole source aquifer protection program established under section 1427;

``(v) the community wellhead protection program established under section 1428;

``(vi) any pesticide or ground water management plan;

``(vii) any voluntary agricultural resource management plan or voluntary whole farm or whole ranch management plan developed and implemented under a process established by the Secretary of Agriculture; and

``(viii) any abandoned well closure program; and

``(C) a description of activities that will be undertaken to coordinate Federal and State programs to respond to the petition.

``(3) <<NOTE: Notification.>> Disapproval.--If the State disapproves a petition submitted under subsection (a), the State shall notify the entity submitting the petition in writing of the reasons for disapproval. A petition may be resubmitted at any time if--

``(A) new information becomes available;

``(B) conditions affecting the source water that is the subject of the petition change; or

``(C) modifications are made in the type of assistance being requested.

``(c) Grants to Support State Programs.--

``(1) In general.--The Administrator may make a grant to each State that establishes a program under this section that is approved under paragraph (2). The amount of each grant shall not exceed 50 percent of the cost of administering the program for the year in which the grant is available.

``(2) Approval.--In order to receive grant assistance under this subsection, a State shall submit to the Administrator for approval a plan for a source water quality protection partnership program that is consistent with the guidance published under subsection (d). The Administrator shall approve the plan if the plan is consistent with the guidance published under subsection (d).

``(d) Guidance.--

``(1) <<NOTE: Publication.>> In general.--Not later than 1 year after the date of enactment of this section, the Administrator, in consultation with the States, shall publish guidance to assist--

``(A) States in the development of a source water quality protection partnership program; and

``(B) municipal or local governments or political subdivisions of a State and community water systems in the development of source water quality protection partnerships and in the assessment of source water quality.

``(2) Contents of the guidance.--The guidance shall, at a minimum--

``(A) recommend procedures for the approval or disapproval by a State of a petition submitted under subsection (a);

``(B) recommend procedures for the submission of petitions developed under subsection (a);

``(C) recommend criteria for the assessment of source water areas within a State; and

``(D) describe technical or financial assistance pursuant to Federal and State programs that is available to address the contamination of sources of drinking water and to develop and respond to petitions submitted under subsection (a).

``(e) Authorization of Appropriations.--There are authorized to be appropriated to carry out this section $5,000,000 for each of the fiscal years 1997 through 2003. Each State with a plan for a program approved under subsection (b) shall receive an equitable portion of the funds available for any fiscal year.

``(f) Statutory Construction.--Nothing in this section--

``(1)(A) creates or conveys new authority to a State, political subdivision of a State, or community water system for any new regulatory measure; or

``(B) limits any authority of a State, political subdivision, or community water system; or

``(2) precludes a community water system, municipal or local government, or political subdivision of a government from locally developing and carrying out a voluntary, incentive-based, source water quality protection partnership to address the origins of drinking water contaminants of public health concern.''.

(b) Sense of the Congress.--It is the sense of the Congress that each State in establishing priorities under section 606(c)(1) of the Federal Water Pollution Control Act should give special consideration to projects that are eligible for funding under that Act and have been recommended pursuant to a petition submitted under section 1454 of the Safe Drinking Water Act.

Sec. 134. Water Conservation Plan

Part E (42 U.S.C. 300j et seq.) is amended by adding at the end the following:

``Sec. 1455. <<NOTE: Federal Register, publication. 42 USC 300j-15.>> (a) Guidelines.--Not later than 2 years after the date of enactment of the Safe Drinking Water Act Amendments of 1996, the Administrator shall publish in the Federal Register guidelines for water conservation plans for public water systems serving fewer than 3,300 persons, public water systems serving between 3,300 and 10,000 persons, and public water systems serving more than 10,000 persons, taking into consideration such factors as water availability and climate.

``(b) Loans or Grants.--Within 1 year after publication of the guidelines under subsection (a), a State exercising primary enforcement responsibility for public water systems may require a public water system, as a condition of receiving a loan or grant from a State loan fund under section 1452, to submit with its application for such loan or grant a water conservation plan consistent with such guidelines.''.

Sec. 135. Drinking Water Assistance to Colonias

Part E (42 U.S.C. 300j et seq.) is amended by adding the following new section at the end thereof:

``Sec. 1456. <<NOTE: 42 USC 300j-16.>> (a) Definitions.--As used in this section:

``(1) Border state.--The term `border State' means Arizona, California, New Mexico, and Texas.

``(2) Eligible community.--The term `eligible community' means a low-income community with economic hardship that--

``(A) is commonly referred to as a colonia;

``(B) is located along the United States-Mexico border (generally in an unincorporated area); and

``(C) lacks a safe drinking water supply or adequate facilities for the provision of safe drinking water for human consumption.

``(b) Grants To Alleviate Health Risks.--The Administrator of the Environmental Protection Agency and the heads of other appropriate Federal agencies are authorized to award grants to a border State to provide assistance to eligible communities to facilitate compliance with national primary drinking water regulations or otherwise significantly further the health protection objectives of this title.

``(c) Use of Funds.--Each grant awarded pursuant to subsection (b) shall be used to provide assistance to one or more eligible communities with respect to which the residents are subject to a significant health risk (as determined by the Administrator or the head of the Federal agency making the grant) attributable to the lack of access to an adequate and affordable drinking water supply system.

``(d) Cost Sharing.--The amount of a grant awarded pursuant to this section shall not exceed 50 percent of the costs of carrying out the project that is the subject of the grant.

``(e) Authorization of Appropriations.--There are authorized to be appropriated to carry out this section $25,000,000 for each of the fiscal years 1997 through 1999.''.

Sec. 136. Estrogenic Substances Screening Program

Part E (42 U.S.C. 300j et seq.) is amended by adding at the end the following:

``Sec. 1457. <<NOTE: 42 USC 300j-17.>> In addition to the substances referred to in section 408(p)(3)(B) of the Federal Food, Drug, and Cosmetic Act (21 U.S.C. 346a(p)(3)(B)) the Administrator may provide for testing under the screening program authorized by section 408(p) of such Act, in accordance with the provisions of section 408(p) of such Act, of any other substance that may be found in sources of drinking water if the Administrator determines that a substantial population may be exposed to such substance.''.

Sec. 137. Drinking Water Studies

Part E (42 U.S.C. 300j et seq.) is amended by adding after section 1457 the following:

``Sec. 1458. <<NOTE: 42 USC 300j-18.>> (a) Subpopulations at Greater Risk.--

``(1) In general.--The Administrator shall conduct a continuing program of studies to identify groups within the general population that may be at greater risk than the general population of adverse health effects from exposure to contaminants in drinking water. The study shall examine whether and to what degree infants, children, pregnant women, the elderly, individuals with a history of serious illness, or other subpopulations that can be identified and characterized are likely to experience elevated health risks, including risks of cancer, from contaminants in drinking water.

``(2) Report.--Not later than 4 years after the date of enactment of this subsection and periodically thereafter as new and significant information becomes available, the Administrator shall report to the Congress on the results of the studies.

``(b) Biological Mechanisms.--The Administrator shall conduct biomedical studies to--

``(1) understand the mechanisms by which chemical contaminants are absorbed, distributed, metabolized, and eliminated from the human body, so as to develop more accurate physiologically based models of the phenomena;

``(2) understand the effects of contaminants and the mechanisms by which the contaminants cause adverse effects (especially noncancer and infectious effects) and the variations in the effects among humans, especially subpopulations at greater risk of adverse effects, and between test animals and humans; and

``(3) develop new approaches to the study of complex mixtures, such as mixtures found in drinking water, especially to determine the prospects for synergistic or antagonistic interactions that may affect the shape of the dose-response relationship of the individual chemicals and microbes, and to examine noncancer endpoints and infectious diseases, and susceptible individuals and subpopulations.

(c) Studies on Harmful Substances in Drinking Water.--

``(1) Development of studies.--The Administrator shall, not later than 180 days after the date of enactment of this section and after consultation with the Secretary of Health and Human Services, the Secretary of Agriculture, and, as appropriate, the heads of other Federal agencies, conduct the studies described in paragraph (2) to support the development and implementation of the most current version of each of the following:

``(A) Enhanced Surface Water Treatment Rule (59 Fed. Reg. 38832 (July 29, 1994)).

``(B) Disinfectant and Disinfection Byproducts Rule (59 Fed. Reg. 38668 (July 29, 1994)).

``(C) Ground Water Disinfection Rule (availability of draft summary announced at (57 Fed. Reg. 33960; July 31, 1992)).

``(2) Contents of studies.--The studies required by paragraph (1) shall include, at a minimum, each of the following:

``(A) Toxicological studies and, if warranted, epidemiological studies to determine what levels of exposure from disinfectants and disinfection byproducts, if any, may be associated with developmental and birth defects and other potential toxic end points.

``(B) Toxicological studies and, if warranted, epidemiological studies to quantify the carcinogenic potential from exposure to disinfection byproducts resulting from different disinfectants.

``(C) The development of dose-response curves for pathogens, including cryptosporidium and the Norwalk virus.

``(3) Authorization of appropriations.--There are authorized to be appropriated to carry out this subsection $12,500,000 for each of fiscal years 1997 through 2003.

``(d) Waterborne Disease Occurrence Study.--

``(1) System.--The Director of the Centers for Disease Control and Prevention, and the Administrator shall jointly--

``(A) within 2 years after the date of enactment of this section, conduct pilot waterborne disease occurrence studies for at least 5 major United States communities or public water systems; and

``(B) <<NOTE: Reports.>> within 5 years after the date of enactment of this section, prepare a report on the findings of the pilot studies, and a national estimate of waterborne disease occurrence.

``(2) Training and education.--The Director and Administrator shall jointly establish a national health care provider training and public education campaign to inform both the professional health care provider community and the general public about waterborne disease and the symptoms that may be caused by infectious agents, including microbial contaminants. In developing such a campaign, they shall seek comment from interested groups and individuals, including scientists, physicians, State and local governments, environmental groups, public water systems, and vulnerable populations.

``(3) Funding.--There are authorized to be appropriated for each of the fiscal years 1997 through 2001, $3,000,000 to carry out this subsection. To the extent funds under this subsection are not fully appropriated, the Administrator may use not more than $2,000,000 of the funds from amounts reserved under section 1452(n) for health effects studies for purposes of this subsection. The Administrator may transfer a portion of such funds to the Centers for Disease Control and Prevention for such purposes.''.

TITLE II--DRINKING WATER RESEARCH

Sec. 201. Drinking Water Research Authorization

Other than amounts authorized to be appropriated to the Administrator of the Environmental Protection Agency under other titles of this Act, there are authorized to be appropriated such additional sums as may be necessary for drinking water research for fiscal years 1997 through 2003. The annual total of such additional sums authorized to be appropriated under this section shall not exceed $26,593,000.

Sec. 202. <<Note: 42 USC 300j-1 Note.>> Scientific Research Review

(a) In General.--The Administrator shall--

(1) develop a strategic plan for drinking water research activities throughout the Environmental Protection Agency (in this section referred to as the ``Agency'');

(2) integrate that strategic plan into ongoing Agency planning activities; and

(3) review all Agency drinking water research to ensure the research--

(A) is of high quality; and

(B) does not duplicate any other research being conducted by the Agency.

(b) <<NOTE: Public information.>> Plan.--The Administrator shall transmit the plan to the Committees on Commerce and Science of the House of Representatives and the Committee on Environment and Public Works of the Senate and the plan shall be made available to the public.

Sec. 203. <<Note: 42 USC 300j-1 Note.>> National Center for Ground Water Research

The Administrator of the Environmental Protection Agency, acting through the Robert S. Kerr Environmental Research Laboratory, is authorized to reestablish a partnership between the Laboratory and the National Center for Ground Water Research, a university consortium, to conduct research, training, and technology transfer for ground water quality protection and restoration. No funds are authorized by this section.

Title III--Miscellaneous Provisions

Sec. 301. Water Return Flows

Section 3013 of Public Law 102-486 (42 U.S.C. 13551) is repealed.

Sec. 302. <<Note: 42 USC 300j-12 Note.>> Transfer of Funds

(a) In General.--Notwithstanding any other provision of law, at any time after the date 1 year after a State establishes a State loan fund pursuant to section 1452 of the Safe Drinking Water Act but prior to fiscal year 2002, a Governor of the State may--

(1) reserve up to 33 percent of a capitalization grant made pursuant to such section 1452 and add the funds reserved to any funds provided to the

State pursuant to section 601 of the Federal Water Pollution Control Act (33 U.S.C. 1381); and

(2) reserve in any year a dollar amount up to the dollar amount that may be reserved under paragraph (1) for that year from capitalization grants made pursuant to section 601 of such Act (33 U.S.C. 1381) and add the reserved funds to any funds provided to the State pursuant to section 1452 of the Safe Drinking Water Act.

(b) Report.--Not later than 4 years after the date of enactment of this Act, the Administrator shall submit a report to the Congress regarding the implementation of this section, together with the Administrator's recommendations, if any, for modifications or improvement.

(c) State Match.--Funds reserved pursuant to this section shall not be considered to be a State match of a capitalization grant required pursuant to section 1452 of the Safe Drinking Water Act or the Federal Water Pollution Control Act (33 U.S.C. 1251 et seq.).

Sec. 303. <<Note: 33 USC 1263a.>> Grants to Alaska to Improve Sanitation in Rural and Native Villages

(a) In General.--The Administrator of the Environmental Protection Agency may make grants to the State of Alaska for the benefit of rural and Native villages in Alaska to pay the Federal share of the cost of--

(1) the development and construction of public water systems and wastewater systems to improve the health and sanitation conditions in the villages; and

(2) training, technical assistance, and educational programs relating to the operation and management of sanitation services in rural and Native villages.

(b) Federal Share.--The Federal share of the cost of the activities described in subsection (a) shall be 50 percent.

(c) Administrative Expenses.--The State of Alaska may use an amount not to exceed 4 percent of any grant made available under this subsection for administrative expenses necessary to carry out the activities described in subsection (a).

(d) Consultation With the State of Alaska.--The Administrator shall consult with the State of Alaska on a method of prioritizing the allocation of grants under subsection (a) according to the needs of, and relative health and sanitation conditions in, each eligible village.

(e) Authorization of Appropriations.--There are authorized to be appropriated $15,000,000 for each of the fiscal years 1997 through 2000 to carry out this section.

Sec. 304. Sense of the Congress

It is the sense of the Congress that appropriations for grants under section 128 (relating to New York City watershed), section 135 (relating to colonias), and section 307 (relating to Alaska Native villages) should not be provided if such appropriations would prevent the adequate capitalization of State revolving loan funds.

Sec. 305. Bottled Drinking Water Standards

Section 410 of the Federal Food, Drug, and Cosmetic Act (21 U.S.C. 349) is amended as follows:

(1) By striking ``Whenever'' and inserting ``(a) Except as provided in subsection (b), whenever''.

(2) By adding at the end the following new subsection:

``(b)(1) Not later than 180 days before the effective date of a national primary drinking water regulation promulgated by the Administrator of the Environmental Protection Agency for a contaminant under section 1412 of the Safe Drinking Water Act (42 U.S.C. 300g-1), the Secretary shall promulgate a standard of quality regulation under this subsection for that contaminant in bottled water or make a finding that such a regulation is not necessary to protect the public health because the contaminant is contained in water in public water systems (as defined under section 1401(4) of such Act (42 U.S.C. 300f(4))) but not in water used for bottled drinking water. <<NOTE: Effective date.>> The effective date for any such standard of quality regulation shall be the same as the effective date for such national primary drinking water regulation, except for any standard of quality of regulation promulgated by the Secretary before the date of enactment of the Safe Drinking Water Act Amendments of 1996 for which (as of such date of enactment) an effective date had not been established. In the case of a standard of quality regulation to which such exception applies, the Secretary shall promulgate monitoring requirements for the contaminants covered by the regulation not later than 2 years after such date of enactment.

``(2) A regulation issued by the Secretary as provided in this subsection shall include any monitoring requirements that the Secretary determines appropriate for bottled water.

``(3) A regulation issued by the Secretary as provided in this subsection shall require the following:

``(A) In the case of contaminants for which a maximum contaminant level is established in a national primary drinking water regulation under section 1412 of the Safe Drinking Water Act (42 U.S.C. 300g-1), the regulation under this subsection shall establish a maximum contaminant level for the contaminant in bottled water which is no less stringent than the maximum contaminant level provided in the national primary drinking water regulation.

``(B) In the case of contaminants for which a treatment technique is established in a national primary drinking water regulation under section 1412 of the Safe Drinking Water Act (42 U.S.C. 300g-1), the regulation under this subsection shall require that bottled water be subject to requirements no less protective of the public health than those applicable to water provided by public water systems using the treatment technique required by the national primary drinking water regulation.

``(4)(A) If the Secretary does not promulgate a regulation under this subsection within the period described in paragraph (1), the national primary drinking water regulation referred to in paragraph (1) shall be considered, as of the date on which the Secretary is required to establish a regulation under paragraph (1), as the regulation applicable under this subsection to bottled water.

``(B) <<NOTE: Federal Register, publication.>> In the case of a national primary drinking water regulation that pursuant to subparagraph (A) is considered to be a standard of quality regulation, the Secretary shall, not later than the applicable date referred to in such subparagraph, publish in the Federal Register a notice--

``(i) specifying the contents of such regulation, including monitoring requirements; and

``(ii) <<NOTE: Effective date.>> providing that for purposes of this paragraph the effective date for such regulation is the same as the effective date for the regulation for purposes of the Safe Drinking Water Act (or, if the exception under paragraph (1) applies to the regulation, that the effective date for the regulation is not later than 2 years and 180 days after the date of enactment of the Safe Drinking Water Act Amendments of 1996).''.

Sec. 306. <<Note: District of Columbia. Virginia. 40 USC 45 note.>> Washington Aqueduct

(a) Definitions.--In this section:

(1) Non-federal public water supply customer.--The terms ``non-Federal public water supply customer'' and ``customer'' mean--

(A) the District of Columbia;

(B) Arlington County, Virginia; and

(C) the city of Falls Church, Virginia.

(2) Secretary.--The term ``Secretary'' means the Secretary of the Army, acting through the Chief of Engineers.

(3) Value to the government.--The term ``value to the Government'' means the net present value of a contract entered into under subsection (e)(2), calculated in accordance with subparagraphs (A) and (B) of section 502(5) of the Congressional Budget Act of 1974 (2 U.S.C. 66la(5)), other than section 502(5)(B)(I) of the Act, as though the contract provided for repayment of a direct loan to a customer.

(4) Washington aqueduct.--The term ``Washington Aqueduct'' means the Washington Aqueduct facilities and related facilities owned by the Federal Government as of the date of enactment of this Act, including--

(A) the dams, intake works, conduits, and pump stations that capture and transport raw water from the Potomac River to the Dalecarlia Reservoir;

(B) the infrastructure and appurtenances used to treat water taken from the Potomac River to potable standards; and

(C) related water distribution facilities.

(b) Regional Entity.--

(1) In general.--The Congress encourages and grants consent to the customers to establish a non-Federal public or private entity, or to enter into an agreement with an existing non-Federal public or private entity, to--

(A) receive title to the Washington Aqueduct; and

(B) operate, maintain, and manage the Washington Aqueduct in a manner that adequately represents all interests of its customers.

(2) Consideration.--If an entity receiving title to the Washington Aqueduct is not composed entirely of non-Federal public water supply customers, the entity shall consider the customers' historical provision of equity for the Aqueduct.

(3) Priority access.--The customers shall have priority access to any water produced by the Washington Aqueduct.

(4) Consent of the congress.--The Congress grants consent to the customers to enter into any interstate agreement or compact required to carry out this section.

(5) Statutory construction.--This section shall not preclude the customers from pursuing any option regarding ownership, operation, maintenance, and management of the Washington Aqueduct.

(c) Progress Report and Plan.--Not later than 1 year after the date of enactment of this Act, the Secretary shall report to the Committee on Environment and Public Works of the Senate and the Committee on Transportation and Infrastructure of the House of Representatives on any progress in achieving the objectives of subsection (b)(1) and shall submit a plan for the transfer of ownership, operation, maintenance, and management of the Washington Aqueduct to a non-Federal public or private entity. Such plan shall include a detailed consideration of any proposal to transfer such ownership, maintenance, or management to a private entity.

(d) Transfer.--

(1) In general.--Subject to subsection (b)(2), the other provisions of this subsection, and any other terms and conditions the Secretary considers appropriate to protect the interests of the United States, the Secretary shall, not later than 3 years after the date of enactment of this Act and with the consent of a majority of the customers and without consideration to the Federal Government, transfer all right, title, and interest of the United States in the Washington Aqueduct, and its real property, facilities, and personalty, to a non-Federal, public or private entity. Approval of such transfer shall not be unreasonably withheld by the Secretary.

(2) Adequate capabilities.--The Secretary shall transfer ownership of the Washington Aqueduct under paragraph (1) only if the Secretary determines, after opportunity for public input, that the entity to receive ownership of the Aqueduct has the technical, managerial, and financial capability to operate, maintain, and manage the Aqueduct.

(3) Responsibilities.--The Secretary shall not transfer title under this subsection unless the entity to receive title assumes full responsibility for performing and financing the operation, maintenance, repair, replacement, rehabilitation, and necessary capital improvements of the Washington Aqueduct so as to ensure the continued operation of the Washington Aqueduct consistent with the Aqueduct's intended purpose of providing an uninterrupted supply of potable water sufficient to meet the current and future needs of the Aqueduct's service area.

(e) Borrowing Authority.--

(1) Borrowing.--

(A) In general.--Subject to the other provisions of this paragraph and paragraph (2), the Secretary is authorized to borrow from the Treasury of the United States such amounts for fiscal years 1997, 1998, and 1999 as are sufficient to cover any obligations that the Army Corps of Engineers is required to incur in carrying out capital improvements during fiscal years 1997, 1998, and 1999 for the Washington Aqueduct to ensure continued operation of the Aqueduct until such time as a transfer of title to the Aqueduct has taken place.

(E) Limitation.--The amount borrowed by the Secretary under subparagraph (A) may not exceed $29,000,000 for fiscal year 1997, $24,000,000 for fiscal year 1998, and $22,000,000 for fiscal year 1999.

(C) Agreement.--Amounts borrowed under subparagraph (A) may only be used for capital improvements agreed to by the Army Corps of Engineers and the customers.

(D) Terms of borrowing.--

(i) In general.--The Secretary of the Treasury shall provide the funds borrowed under subparagraph (A) under such terms and conditions as the Secretary of Treasury determines to be necessary and in the public interest and subject to the contracts required under paragraph (2).

(ii) Term.--The term of any loan made under subparagraph (A) shall be for a period of not less than 20 years.

(iii) Prepayment.--There shall be no penalty for the prepayment of any amounts borrowed under subparagraph (A).

(2) Contracts with customers.--

(A) In general.--The borrowing authority under paragraph (1)(A) shall be effective only after the Chief of Engineers has entered into contracts with each customer under which the customer commits to repay a pro rata share (based on water purchase) of the principal and interest owed by the Secretary to the Secretary of the Treasury under paragraph (1).

(B) Prepayment.--Any customer may repay, at any time, the pro rata share of the principal and interest then owed by the customer and outstanding, or any portion thereof, without penalty.

(C) Risk of default.--Under each of the contracts, the customer that enters into the contract shall commit to pay any additional amount necessary to fully offset the risk of default on the contract.

(D) Obligations.--Each contract under subparagraph (A) shall include such terms and conditions as the Secretary of the Treasury may require so that the value to the Government of the contracts entered into under subparagraph (A) is estimated to be equal to the obligations of the

Army Corps of Engineers for carrying out capital improvements at the Washington Aqueduct at the time that each series of contracts is entered into.

(E) Other conditions.--Each contract entered into under subparagraph (A) shall--

(i) provide that the customer pledges future income only from fees assessed for principal and interest payments required by such contracts and costs to operate and maintain the Washington Aqueduct;

(ii) provide the United States priority in regard to income from fees assessed to operate and maintain the Washington Aqueduct; and

(iii) include other conditions consistent with this section that the Secretary of the Treasury determines to be appropriate.

(3) Limitations.--

(A) Borrowing authority.--The Secretary's borrowing authority for making capital improvements at the Washington Aqueduct under paragraph (1) shall not extend beyond fiscal year 1999.

(B) Obligation authority.--Upon expiration of the borrowing authority exercised under paragraph (1), the Secretary shall not obligate funds for making capital improvements at the Washington Aqueduct except funds which are provided in advance by the customers. This limitation does not affect the Secretary's authority to conduct normal operation and maintenance activities, including minor repair and replacement work.

(4) <<NOTE: Reports.>> Impact on improvement program.--Not later than 180 days after the date of enactment of this Act, the Secretary, in consultation with other Federal agencies, shall transmit to the Committee on Environment and Public Works of the Senate and the Committee on Transportation and Infrastructure of the House of Representatives a report that assesses the impact of the borrowing authority provided under this subsection on the near-term improvement projects in the Washington Aqueduct Improvement Program, work scheduled, and the financial liability to be incurred.

(f) Reissuance of NPDES Permit.--Prior to reissuing a National Pollutant Discharge Elimination System (NPDES) permit for the Washington Aqueduct, the Administrator of the Environmental Protection Agency shall consult with the customers and the Secretary regarding opportunities for more efficient water facility configurations that might be achieved through various possible transfers of the Washington Aqueduct.

Such consultation shall include specific consideration of concerns regarding a proposed solids recovery facility, and may include a public hearing.

Sec. 307. <<Note: 33 USC 1281 Note.>> Wastewater Assistance to Colonias

(a) Definitions.--As used in this section:

(1) Border state.--The term ``border State'' means Arizona, California, New Mexico, and Texas.

(2) Eligible community.--The term ``eligible community'' means a low-income community with economic hardship that--

(A) is commonly referred to as a colonia;

(B) is located along the United States-Mexico border (generally in an unincorporated area); and

(C) lacks basic sanitation facilities such as household plumbing or a proper sewage disposal system.

(3) Treatment works.--The term ``treatment works'' has the meaning provided in section 212(2) of the Federal Water Pollution Control Act (33 U.S.C. 1292(2)).

(b) Grants for Wastewater Assistance.--The Administrator of the Environmental Protection Agency and the heads of other appropriate Federal agencies are authorized to award grants to a border State to provide assistance to eligible communities for the planning, design, and construction or improvement of sewers, treatment works, and appropriate connections for wastewater treatment.

(c) Use of Funds.--Each grant awarded pursuant to subsection (b) shall be used to provide assistance to one or more eligible communities with respect to which the residents are subject to a significant health risk (as determined by the Administrator or the head of the Federal agency making the grant) attributable to the lack of access to an adequate and affordable treatment works for wastewater.

(d) Cost Sharing.--The amount of a grant awarded pursuant to this section shall not exceed 50 percent of the costs of carrying out the project that is the subject of the grant.

(e) Authorization of Appropriations.--There are authorized to be appropriated to carry out this section $25,000,000 for each of the fiscal years 1997 through 1999.

Sec. 308. Prevention and Control of Zebra Mussel Infestation of Lake Champlain

(a) Findings.--Section 1002(a) of the Nonindigenous Aquatic Nuisance Prevention and Control Act of 1990 (16 U.S.C. 4701(a)) is amended as follows:

(1) By striking ``and" at the end of paragraph (3).

(2) By striking the period at the end of paragraph (4) and inserting ``; and".

(3) By adding at the end the following new paragraph:

``(5) the zebra mussel was discovered on Lake Champlain during 1993 and the opportunity exists to act quickly to establish zebra mussel controls before Lake Champlain is further infested and management costs escalate.".

(b) Ex Officio Members of Aquatic Nuisance Species Task Force.--

Section 1201(c) of such Act (16 U.S.C. 4721(c)) is amended by inserting ``, the Lake Champlain Basin Program," after ``Great Lakes Commission".

Title IV--Additional Assistance for Water Infrastructure and Watersheds

Sec. 401. <<Note: 42 USC 300j-3c.>> National Program

(a) Technical and Financial Assistance.--The Administrator of the Environmental Protection Agency may provide technical and financial assistance in the form of grants to States (1) for the construction, rehabilitation, and improvement of water supply systems, and (2) consistent with nonpoint source management programs established under section 319 of the Federal Water Pollution Control Act, for source water quality protection programs to address pollutants in navigable waters for the purpose of making such waters usable by water supply systems.

(b) Limitation.--Not more than 30 percent of the amounts appropriated to carry out this section in a fiscal year may be used for source water quality protection programs described in subsection (a)(2).

(c) Condition.--As a condition to receiving assistance under this section, a State shall ensure that such assistance is carried out in the most cost-effective manner, as determined by the State.

(d) Authorization of Appropriations.--

(1) Unconditional authorization.--There are authorized to be appropriated to carry out this section $25,000,000 for each of fiscal years 1997 through 2003. Such sums shall remain available until expended.

(2) Conditional authorization.--In addition to amounts authorized under paragraph (1), there are authorized to be appropriated to carry out this title $25,000,000 for each of fiscal years 1997 through 2003, provided that such authorization shall be in effect for a fiscal year only if at least 75 percent of the total amount of funds authorized to be appropriated for such fiscal year by section 1452(m) of the Safe Drinking Water Act are appropriated.

(e) Acquisition of Lands.--Assistance provided with funds made available under this title may be used for the acquisition of lands and other interests in lands; however, nothing in this title authorizes the acquisition of lands or other interests in lands from other than willing sellers.

(f) Federal Share.--The Federal share of the cost of activities for which grants are made under this title shall be 50 percent.

(g) Definitions.--In this section, the following definitions apply:

(1) State.--The term ``State'' means a State, the District of Columbia, the Commonwealth of Puerto Rico, the Virgin Islands, Guam, American Samoa, and the Commonwealth of the Northern Mariana Islands.

(2) Water supply system.--The term ``water supply system'' means a system for the provision to the public of piped water for human consumption if such system has at least 15 service connections or regularly serves at least 25 individuals and a draw and fill system for the provision to the public of water for human consumption. Such term does not include a system owned by a Federal agency. Such term includes (A) any collection, treatment, storage, and distribution facilities under control of the operator of such system and used primarily in connection with such system, and (B) any collection or pretreatment facilities not under such control that are used primarily in connection with such system.

TITLE V--CLERICAL AMENDMENTS

Sec. 501. Clerical Amendments

(a) Part B.--Part B (42 U.S.C. 300g et seq.) is amended as follows:

(1) In section 1412(b), <<NOTE: 42 USC 300g-1.>> move the margins of paragraph (11) 2 ems to the right.

(2) In section 1412(b)(8), strike ``1442(g)" and insert ``1442(e)".

(3) In section 1415(a)(1)(A), <<NOTE: 42 USC 300g-4.>> insert ``the" before ``time the variance is granted".

(b) Part C.--Part C (42 U.S.C. 300h et seq.) is amended as follows:

(1) In section 1421(b)(3)(B)(i), <<NOTE: 42 USC 300h.>> strike ``number or States" and inserting ``number of States".

(2) In section 1427(k), <<NOTE: 42 USC 300h-6.>> strike ``this subsection" and inserting ``this section".

(c) Part E.--Section 1441(f) (42 U.S.C. 300j(f)) is amended by inserting a period at the end.

(d) Section 1465(b).--Section 1465(b) (42 U.S.C. 300j-25(b)) is amended by striking ``as by" and inserting ``by".

(e) Short Title.--Section 1 of Public Law 93-523 (88 Stat. 1600) is <<NOTE: 42 USC 201 note. ``SHORT TITLE>> amended by inserting ``of 1974" after ``Act" the second place it appears and title XIV of the Public Health Service Act is amended by inserting the following immediately before part A:

``Sec. 1400. This title may be cited as the `Safe Drinking Water Act'.".

(f) Technical Amendments to Section Headings.--

(1) The section heading and subsection designation of subsection (a) of section 1417 (42 U.S.C. 300g-6) are amended to read as follows:

``Sec. 1417. (a)".

(2) The section heading and subsection designation of subsection (a) of section 1426 (42 U.S.C. 300h-5) are amended to read as follows:

``Sec. 1426. (a)".

(3) The section heading and subsection designation of subsection (a) of section 1427 (42 U.S.C. 300h-6) are amended to read as follows:

``Sec. 1427. (a)".

(4) The section heading and subsection designation of subsection (a) of section 1428 (42 U.S.C. 300h-7) are amended to read as follows:

``Sec. 1428. (a)".

(5) The section heading and subsection designation of subsection (a) of section 1432 (42 U.S.C. 300i-1) are amended to read as follows:

``Sec. 1432. (a)".

(6) The section heading and subsection designation of subsection (a) of section 1451 (42 U.S.C. 300j-11) are amended to read as follows:

``Sec. 1451. (a)".

(7) The section heading and first word of section 1461 (42 U.S.C. 300j-21) are amended to read as follows:

``Sec. 1461. As''.

(8) The section heading and first word of section 1462 (42 U.S.C. 300j-22) are amended to read as follows:

``Sec. 1462. For''.

(9) The section heading and subsection designation of subsection (a) of section 1463 (42 U.S.C. 300j-23) are amended to read as follows:

``Sec. 1463. (a)''.

(10) The section heading and subsection designation of subsection (a) of section 1464 (42 U.S.C. 300j-24) are amended to read as follows:

``Sec. 1464. (a)''.

(11) The section heading and subsection designation of subsection (a) of section 1465 (42 U.S.C. 300j-25) are amended to read as follows:

``Sec. 1465. (a)''.

Approved August 6, 1996.

LEGISLATIVE HISTORY--S. 1316 (H.R. 3604)

House Reports: ?Nos. 104-632, Pt. 1 accompanying H.R. 3604 (Comm. on Commerce) and 104-741 (Comm. of Conference).

Senate Reports: ?No. 104-169 (Comm. on Environment and Public Works).

Congressional Record:

Vol. 141 (1995):

Nov. 29, considered and passed Senate.

Vol. 142 (1996):

June 25, H.R. 3604 considered and passed House.

July 17, S. 1316 considered and passed House, amended, in lieu of H.R. 3604.

Aug. 2, House and Senate agreed to conference report.

Weekly Compilation of Presidential Documents, Vol. 32 (1996)

Aug. 6, Presidential remarks and statement.

INDEX

A

B

C

D

S

T